SEISMOLOGY

SEISMOLOGY

HUGH DOYLE
Department of Geology and Geophysics
University of Western Australia

JOHN WILEY & SONS
Chichester · New York · Brisbane · Toronto · Singapore

National 01243 779777
International (+44) 1243 779777

Other Wiley Editorial Offices

John Wiley & Sons, Inc., 605 Third Avenue,
New York, NY 10158-0012, USA

Jacaranda Wiley Ltd, 33 Park Road, Milton,
Queensland 4064, Australia

John Wiley & Sons (Canada) Ltd, 22 Worcester Road,
Rexdale, Ontario M9W 1L1, Canada

John Wiley & Sons (SEA) Pte Ltd, 37 Jalan Pemimpin #05-04,
Block B, Union Industrial Building, Singapore 2057

Library of Congress Cataloging-in-Publication Data

Doyle, H.A. (Hugh A.)
Seismology / Hugh Doyle.
p. cm.
Includes bibliographical references and index.
ISBN 0-471-94869-1 (alk. paper)
1. Seismology. I. Title.
QE534.2.D69 1995
551.2′2—dc20

95–9101
CIP

British Library Cataloguing in Publication Data

A catalogue record for this book is available from the British Library

ISBN 0 471 94869 1

Typeset in 10/12pt Times by Vision Typesetting, Manchester
Printed and bound in Great Britain by Redwood Books, Trowbridge
This book is printed on acid-free paper responsibly manufactured from sustainable forestation, for which at least two trees are planted for each one used for paper production.

Contents

Preface

Seismology is an important branch of earth science and geophysics, providing most of our knowledge of the structure of the Earth. Seismology is used in various investigations of the sub-surface, being essential in modern exploration for oil and gas, in which most seismologists are employed. The study of earthquakes is a fascinating subject both for the professional and the layman, and has increasing importance as populations grow and spread. Seismology continues to grow and become more sophisticated with the development of more and better instruments and surveys and the impact of the computer.

Usually texts on the subject cover either exploration methods or general seismology only. I thought it worthwhile to survey the whole field of the subject from general seismology, earthquakes, Earth structure to exploration methods. Mathematics is kept to a minimum but with plenty of illustrations to make clear the principles. It is aimed at students of geophysics and geology and other earth scientists and professionals.

I would like to acknowledge the important contributions of Colin Steel who redrew most of the figures and of Iain Stevenson of John Wiley & Sons. The following people read parts of the text: Drs Michael Dentith, Ron List and Carl Knox-Robinson of the University of Western Australia, Peter Vaughan of Woodside Petroleum and Terry Allen of Western Geophysical.

The following people kindly provided figures or permission to use them: Dr George Plafker, Michael Moore, Mary Lou Zoback and Dr R. E. Wallace, US Geological Survey, Menlo Park, California, USA; Henry Spall, US Geological Survey, Reston, Virginia, USA; Sharon Anker, Australian Geological Survey Organisation, Canberra, Australia; Soraya Brombacher, Western Atlas Int., Houston, Texas, USA; Victor Dent, Mundaring Observatory; Kevin McCue, David Denham and the Director of the Australian Geological Survey Organisation, Canberra, Australia; Richard Goto, Geco-Prakla, Gatwick, UK; Dr A. Hasegawa, Tohoku University, Sendai, Japan; Dr Akira Matsuzawa, Japan China Oil Dev. Corp., Tokyo, Japan; Tony Macpherson, Geco-Prakla, Brendale, Queensland, Australia; Pat Orr, California Institute of Technology, USA; R. Purcell, P & R Geological Consultants, Perth, Australia; Edwin S. Robinson, Virginia

Polytechnical Institute, USA; Bob Sheriff, University of Houston, Texas, USA; Andrew Sutherland, Geco-Prakla, Melbourne, Australia; and Sean Waddingham, Horizon Exploration, Sevenoaks, UK.

Glossary

Accelerometer A seismograph designed to measure ground acceleration.

Alluvial Unconsolidated terrestrial sediment.

Analogue A continuous physical variable such as a voltage (in contrast to digital).

Anelasticity Not 'elastic', i.e. strain not proportional to stress.

Angle of incidence The angle between an incident ray and the perpendicular to the interface, and between the wavefront and interface.

Anticline A fold in stratified rocks, convex upward.

Band-width The range of frequencies.

Basalt A fine-grained dark mafic rock, usually volcanic.

Basic rock An igneous rock with mafic minerals, rich in iron and magnesium.

Batholith A large irregular mass of igneous rock, intruded into the upper crust.

Bedding Stratification of sedimentary rock.

Caldera A large circular depression, volcanic in origin.

Clastic A sedimentary rock, largely made up of fragments, e.g. sandstone and conglomerates.

Coherence A measure of the similarity of two functions, e.g. wave trains.

Continental shelf The gently sloping, submerged edge of a continent.

Craton A large, ancient, stable part of the Earth's crust.

Critical angle The angle of incidence for which the refracted ray grazes the interface between two layers.

Dextral In a dextral (right-lateral) fault the opposite side moved to the right.

Dilatation An increase (usually slight) in volume; opposite of compression.

Ductile Capable of considerable change of shape without fracture.

Dynamic range The ability to record large and small amplitude signals.

Earth tides Deformation of the solid Earth by tidal forces.

Evaporite A sediment produced by evaporation.

Faulting The displacement of rocks along a shear surface.

Fault creep A slow gradual deformation of a fault in contrast to fracture.

Fault plane The plane that best approximates the fracture surface.

Felsic rock A light-coloured igneous rock abundant in feldspar and quartz and poor in iron and magnesium, in contrast to basic rock.

Floating point A number expressed as a power, often binary, e.g. 2.51×2^7.

Fluid pore-pressure The pressure within the fluid of a porous rock.

Fractal A fractal process is one in which the spatial pattern is the same at all scales, often following a power law.

Gabbro A coarse-grained dark rock; the deeper, intrusive equivalent of basalt.

Gather A group of seismic traces which have a co-ordinate in common, e.g. a common-mid-point gather.

Geoid The Earth's shape as given by the gravity equipotential surface.

Geosyncline A major downwarp in the crust with deep sediments.

Granulite A coarse-grained metamorphic rock generally formed at high temperature and pressure.

Gravimeter An instrument which measures the strength of gravity.

Half-space A mathematical model bounded by only one plane surface.

Harmonic A frequency which is a multiple of a fundamental frequency.

Head wave The wave which is critically refracted at an interface.

High-pass filter A filter which passes frequencies above a cut-off frequency.

Hot spot A warmer region of the Earth produced by a convective plume; e.g. Hawaii and Iceland.

Huyghen's principle Every point on a wave-front is a source of secondary waves and later wave-fronts are envelopes of the secondaries.

Igneous intrusion Penetration of rock by a body of magma.

Incident angle The angle that a ray makes with the perpendicular to an interface; also that of a wave-front to the interface.

Inversion Calculation of model values from observed data, with certain assumptions and limits for the model.

Island arc A linear or arcuate chain of volcanic islands formed by plate subduction.

Isostasy, isostatic The concept that areas of the crust are in gravitational balance, mountainous areas being underlain by thicker, low density crust, and low and oceanic areas by thinner crust.

Isotropic Having the same properties in all directions.

JOIDES Joint Oceanographic Institutions for Deep Earth Sampling.

Kirchhoff's equation An integral form of the wave equation giving the wave function in terms of earlier values and their derivatives.

Least-squares filter (*Wiener filter*) Changes an input into near a desired output, such that the sum of the squares of the differences between the output and the input is minimized.

Left-lateral In a left-lateral fault (sinistral) the opposite side moved to the left.

Liquefaction Loss of strength of saturated soil or sand from strong vibration. Compaction releases water and the soil may flow.

Mafic rock Igneous rock with low silica content and high iron–magnesium content.

Magma Molten igneous rocks; lava on reaching the surface.

Metamorphic grade The relative amount of change in mineralogy and texture produced in the metamorphism by temperature and pressure.

Mid-ocean ridge The major belt of raised ocean floor where the ocean plates diverge and are created by upwelling magma.

Mode In seismic waves or oscillations, a particular type of vibration.

Modulus of elasticity An elastic constant, the ratio of stress to strain.

Moment The turning force or torque, force times radius.

Multifold In CMP recording the use of many reflections from the same mid-point.

Multiplexing The process which converts many channels of data into one by sampling channels in turn.

Muting The exclusion of the early part of a reflection record where it is confused by noise and refractions.

Near field The field near the source of radiation.

Negative feedback Feeding a small part of an amplifier's output back into the input after reversing its phase.

Obduction The pushing of a piece of subducted plate (ophiolites) onto an overriding plate during plate collision.

Ophiolites A group of obducted oceanic crustal rocks.

Outcrop A portion of exposed bedrock.

Overburden Unconsolidated material over bedrock.

Peg-leg multiples Short-path multiple reflections between thin beds.

Peridotite A coarse-grained mafic igneous rock.

Phase change The change of a solid mineral to a different structure and density.

Phase velocity The velocity of a particular part of a wave, such as the peak or trough.

Piezoelectricity The property whereby a crystal produces a small voltage when stressed, e.g. by pressure.

Pillow lavas Pillow-shaped extrusions of lava formed underwater.

Plane-polarized The vibration of a wave confined to a plane.

Plate tectonics The theory of the Earth as divided into a number of lithospheric plates moving relatively to each other by convection below.

Plutonic Igneous activity at depth.

Porosity The percentage of pore space in a rock.

Power spectral density Ground motion energy per unit time as a function of frequency.

Profile A series of recordings along a line, or an ensemble of such data.

Rarefaction A depression, a dilatation.

Reef A local limestone structure, e.g. of coral.

Regolith A general term for a surface layer of fragmental, loose or unconsolidated rock and soil.
Resolution The ability to separate two close features.
Retrograde Rotation in opposite direction, e.g. in a Rayleigh wave relative to a water wave.
Reverse fault One in which the upper block moves up and over the lower block.
Right-lateral The other side of a fault moved to the right (dextral).
Section A plot of seismic data along a traverse, or a slice through a model.
Seismicity The frequency, location and magnitude of earthquakes.
Seismic sequence A sequence of sedimentary layers defined by seismic data, normally using unconformities to limit a sequence.
Seismoscope An early device that recorded only that a vibration had occurred without producing a seismogram.
Semblance A measure of the coherence between seismic channels.
Semi-infinite Extending to infinity in one or two directions.
Sequence analysis The grouping of sedimentary layers in reflection data, particularly with the use of unconformities.
Shear A displacement in which the planes on either side remain parallel.
Shield A large region of ancient, stable, basement rocks in a continent.
Shoran A short-range radio ranging system using timed pulses.
Shot-time The time of an explosion or other seismic source.
Sign-bit The bit indicating the sign (plus or minus) of a number.
Slip The relative motion across a fault.
Spheroidal mode The type of Earth oscillation in which the globe takes various spheroidal shapes, e.g. one axis longer than the other.
Stacking Combining different seismic records, e.g. those with common mid-point (CMP stack).
Standing wave The result of the interference of two waves travelling in different, usually opposite, directions.
Stratigraphy The study of the correlation and sequence of sedimentary rocks.
Strike-slip Slip along the direction of strike (outcrop), thus horizontal.
Subduction zone Where a sinking plate is moving slowly beneath an over-riding plate.
Tectonic Relating to Earth structures, usually the major ones.
Teleseism A distant earthquake.
Thrust fault A reverse fault, especially with a dip of less than 45°.
Tiltmeter A device for measuring ground tilt; as in solid Earth tides.
Time domain Where a variable is given as a function of time.
Toroidal mode An oscillation in which the body twists, no radial motion.
Transcurrent fault A strike-slip or wrench fault.
Transducer A device which transforms one type of energy to another, e.g. motion to electrical energy, as in a seismometer.

Transform fault A fault displacing rifts such as the mid-ocean rifts.

Transgression A landward movement of the shoreline.

Traverse A survey line.

Unconformity A surface which has been eroded or had no deposition of sediments.

Underdamped A lack of damping causing, for example, a seismograph to resonate.

Variable-area Seismic data shown with black areas approximately proportional to amplitude.

Velocity analysis Calculation of stacking or NMO velocity.

Wave equation The mathematical relation between space and time for a wave.

Wave-front The surface perpendicular to a ray. It has a certain phase.

Weathering The shallow low-velocity layer of weathered rocks.

Wiener filter A filter which changes an input to near a desired output by minimizing the differences between input and output in a least-squares sense.

Chapter 1
Introduction

1.1 INTRODUCTION

If we wish to look into the Earth's interior, the most accurate method is that of seismology, measuring the elastic earth waves passing through the globe or reflected or refracted back to the surface. The word 'seismology' comes from the Greek *seismos* (earthquake) but today it refers to the branch of geophysics concerned not only with earthquakes, but with the recording and analysis of any vibration in the Earth (and Moon or planets) produced naturally or artificially. This includes vibrations through volcanism, and also the continual background vibration (microseisms) caused by ocean and atmospheric waves and wind.

We can summarize the branches of seismology as

(1) earthquake recording and the determination of their location, magnitude, moment, energy and fault motion (Chapters 3–9);
(2) the study of Earth structure and that of other planets using seismic waves to obtain images of the interior (Chapters 10–13);
(3) the imaging of the geology of sedimentary basins in the search for oil, gas and coal and, to a much less extent, imaging 'hardrock' areas in the exploration for metals (Chapters 14–21), measuring ice thicknesses in glaciers, using explosions and other energy sources;
(4) shallow surveys for hydrology, groundwater exploration, and the investigation of foundations of buildings, dams, roads; also using explosions, etc.;
(5) theoretical or mathematical seismology and data processing.

Seismology is the most accurate method of imaging the interior for exploration, particularly for imaging sedimentary basins with their layered structures (Figure 1.1). Fortunately the Earth is transparent to seismic waves (of reasonably low frequency) so that the measurements of travel-times, refractions, reflections and oscillations of the whole Earth have gradually revealed more about its structure than any other method (Figure 1.2). The exploration for oil and gas would be

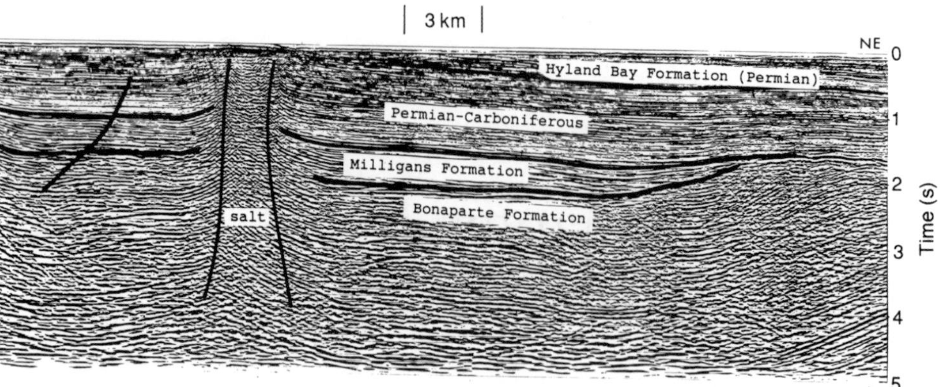

Figure 1.1 Sedimentary layers and a salt dome mapped in a seismic reflection survey, north-west Australia. (Courtesy of R. Purcell and the Australian Petroleum Exploration Association)

almost impossible today without it. Such seismic exploration is largely dependent on the differences in velocity of waves in different rocks or layers. Thus measurements of the travel-times of reflected and refracted waves are the major data source in exploration, although the amplitude and frequency of waves and even absorption also now play a part.

Many books on seismology have been published, but they address either earthquake seismology and the structure of the Earth, or exploration seismology. This book is an attempt to introduce the reader to both aspects of the subject in one volume and with little mathematics. A number of other texts on seismology, including mathematical seismology, are listed in Appendix 1.

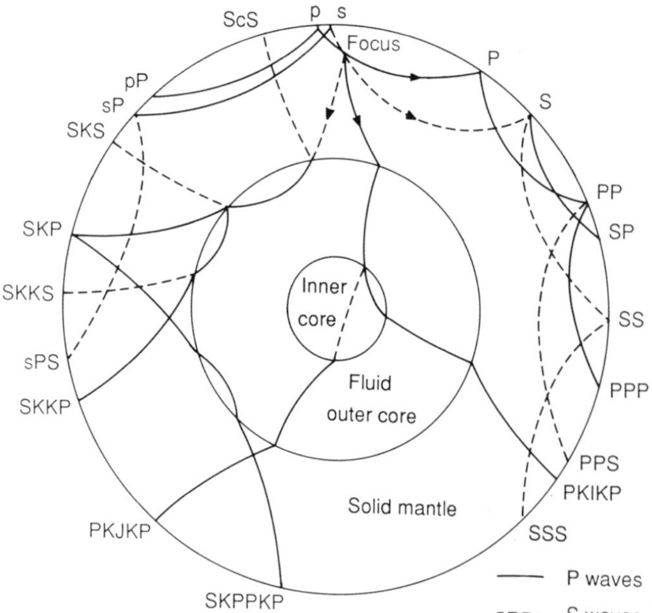

Figure 1.2 Some of the seismic ray paths of refracted and reflected waves through the Earth. A P wave reflected at the surface produces a PP and a PS (converted to S on reflection). Similarly an S wave produces an SS and an SP. The symbols c and i denote reflections from the outer and inner core boundaries respectively, p and s are for P and S arriving directly at the surface and almost above the focus. K and I are P waves travelling through the outer and inner cores, and J for S in the inner core. Thus ScS is an S wave reflected off the core and SKP is one converted to a P-type wave in the outer core and refracted back to the surface.

1.2 BRIEF HISTORY OF SEISMOLOGY

1.2.1 General seismology

Seismology began with the study of earthquakes, and the realization that the transmission of elastic waves was involved, by such as Hooke in 1668 and Michell of Cambridge in his study of earthquakes in 1760. Michell also suggested the possibility of stars large enough to be black-holes and invented a torsion balance which Cavendish used in 1798 to 'weigh the Earth' (actually to determine G and the mean density of the Earth). In the 19th century French and later English mathematicians developed the theory of seismic waves through a solid body. In 1828 Poisson was the first to show that there are two interior ('body') waves, compressive (P) and shear (S), produced by a disturbance in a solid body (Figure 1.3). Surface waves were predicted by Rayleigh in 1887 and both predictions were verified in 1899 by Oldham, who studied the seismograms of the great Indian earthquake of 1897 recorded on the newly developed seismographs. Acceptance of faulting as the main cause of earthquakes came with the study of the 1891 Mino-Awari and 1906 San Francisco earthquakes (Howell, 1986).

Early in the 19th century seismoscopes such as that of Forbes in Scotland (in 1841) were built. These simply used pendulums to trace motion in sand or on smoked glass or paper and were insensitive (Davison, 1927; Stoneley, 1967; Howell, 1989). Horizontal pendulums (as in Figure 1.4 which shows a modern instrument) were later employed by Palmieri in Italy. Milne, Ewing and Gray, working in Japan, developed better instruments, the Milne being used in the first, though limited, world seismic network of 15 stations around the British Empire (Howell, 1986). The Milne instrument was under-damped (making it resonate too much) and had a magnification of only about 8 at most periods. Later, Golitsyn in Russia devised the more sensitive electromagnetic seismograph with photographic recording which became the standard type (see Melton (1981) on developments up to that date) until the recent development of the broad-band and digital seismographs (p. 38).

Oldham (1906) first noticed the seismic evidence for the Earth's

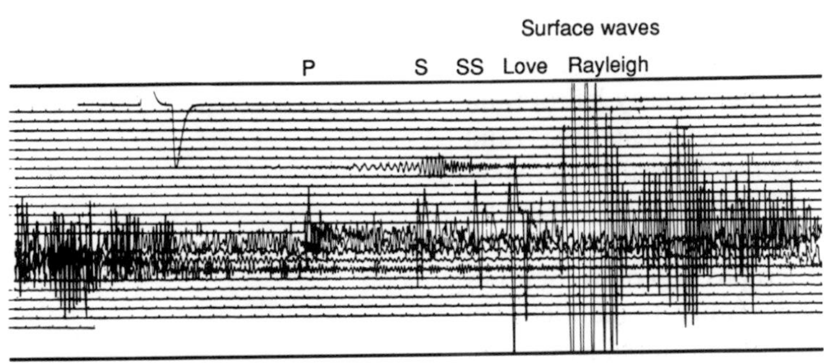

Figure 1.3 A seismogram of a Tongan earthquake recorded at Mundaring, Western Australia. (Courtesy of Director, Aust. Geol. Surv. Org.) Note the P and S waves, and Love and Rayleigh surface waves

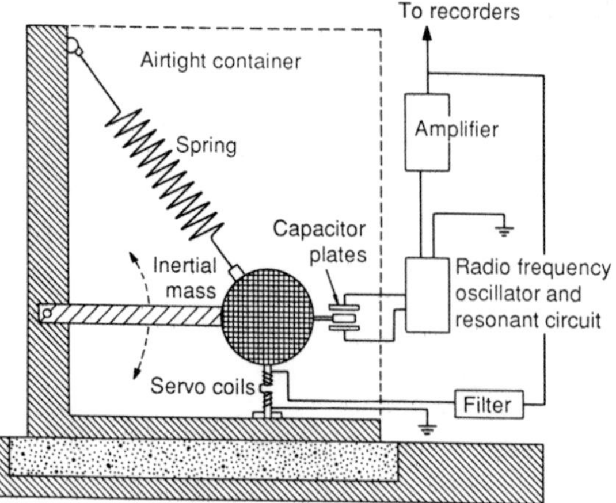

Figure 1.4 A vertical pendulum seismograph. The motion at the capacitor plates produces an electrical signal which is amplified and recorded with appropriate time marks. (After Press, 1965)

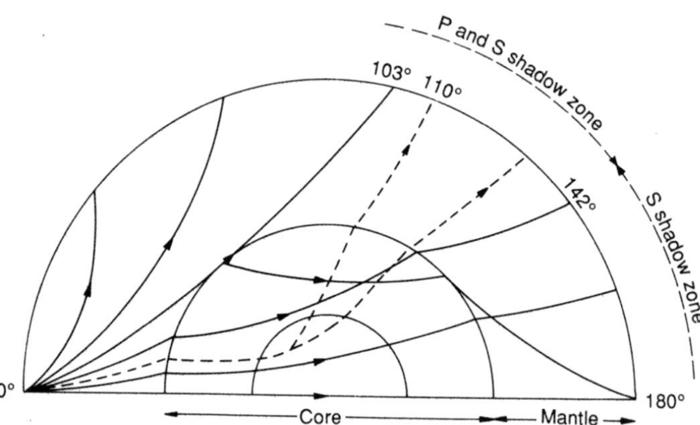

Figure 1.5 The P and S 'shadow' zones produced by the lower velocity core. (After Gutenberg, 1959)

core, namely the shadow zone beyond about 105° distance (Figure 1.5) using early earthquake records (note that large distances are measured by the angle subtended at the centre of the Earth by the seismic station and the earthquake position). In 1914 Gutenberg accurately determined the depth of the core boundary as 2900 km. Jeffreys (1926a) finally showed that the low rigidity of the outer core indicated that it must be liquid, and in 1936 Inge Lehmann published the evidence for the inner core (Brush, 1980).

Important work on seismic travel-times through the Earth was carried out in the 1920s, and particularly in the 1930s by Jeffreys and Bullen (1940), Gutenberg and Richter (also on seismicity, 1954) and Wadati. The latter gave the first good evidence of deep earthquakes on inclined planes (Frohlich, 1987) (see p. 45). Pioneering studies on Earth models were made by Bullen.

The number and the sensitivity of seismographs in various countries slowly increased during the first half of the century. In the 1960s US authorities established a much larger and more advanced network of (now) 125 seismographs, the Worldwide Standard Seismograph Network (WWSN) (Figure 1.6), using Benioff instruments, partly to detect underground nuclear explosions. Finance for seismic research increased greatly with the need to study nuclear explosions (Howell, 1990). The worldwide network greatly improved the accuracy of studies of global seismicity and fault motion and proved important in the development of plate tectonics.

Now the rather obsolete WWSN is being replaced by the Global Digital Seismograph Network (GDSN) with a number of wide-band and digitally recorded instruments (Romanowicz and Dziewonski, 1987; Choy, 1990) and by the new US National Seismograph Network recorded centrally via satellite (plus regional networks).

In 1909 Mohorovičić had noticed evidence in a local earthquake recording for the boundary or discontinuity in velocity (a sudden increase) which bears his name and has been taken as the definition of the base of the crust and the top of the mantle (Figure 1.7). Since then many studies of crustal structure around the world have been carried out using refraction and reflection methods from earthquake and explosive sources. Ewing of the Columbia University, New York, and Hill of Cambridge, UK, pioneered seismic surveys at sea in the late 1930s and 1940s, revealing the shallower oceanic crust and its different geological structure (Figure 1.8).

In the 1960s the development of seismographs of long natural period and computers made possible more detailed calculations and modelling of surface wave dispersion (Dorman *et al.*, 1960). This greatly added to our knowledge of crustal and mantle structure. Later, detection of oscillations of the Earth enabled further refining of whole Earth models, in particular the PREM model of Dziewonski and Anderson (1981). Other important developments depending on the computer were the calculation of synthetic seismograms and 3D models of deep Earth structure, some by tomographic modelling (p. 97).

Seismic data, particularly accurate earthquake locations (seismicity maps) and the calculated directions of fault movement, have been an important part of the information making possible the development of plate tectonics, our understanding of the slow movements of great plates of lithosphere (Figure 1.9) over the Earth's surface at rates of centimetres per year (e.g. see Kearey and Vine, 1990; Gordon and Stein, 1992). The seismicity maps first showed the existence of the globe-girdling mid-ocean ridge system. For histories of general seismology, see Davison (1927), Stoneley (1967), Dewey and Byerly (1969), Howell (1986) and Agnew (1989).

1.2.2 Exploration seismology

Robert Mallett (1848) of Ireland pointed out the use of seismic velocities in investigating the Earth's interior and seems to have been the first to attempt to measure seismic velocities using gunpowder

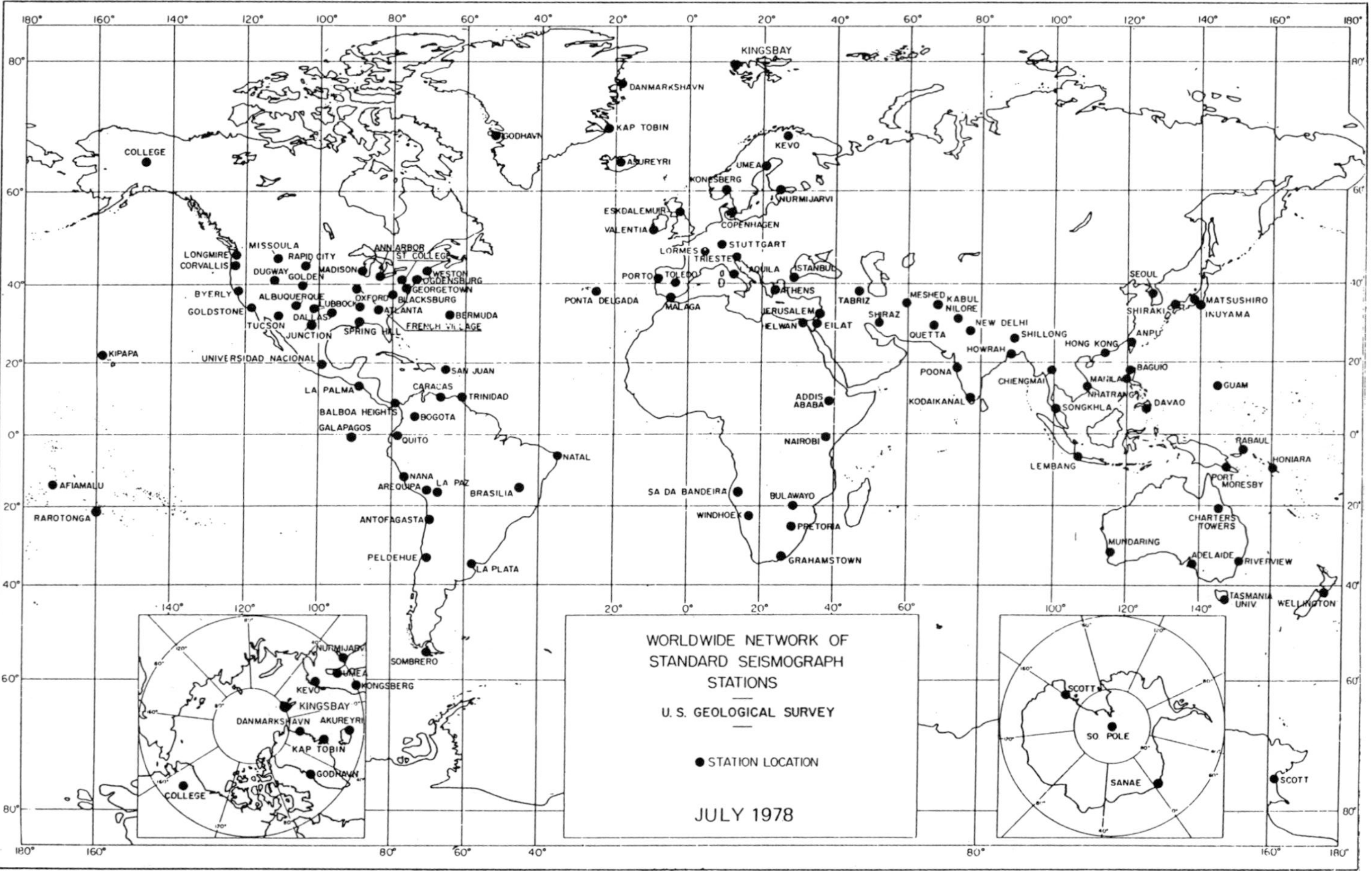

Figure 1.6 Distribution of WWSN seismograph stations around the world. (Courtesy US Geological Survey)

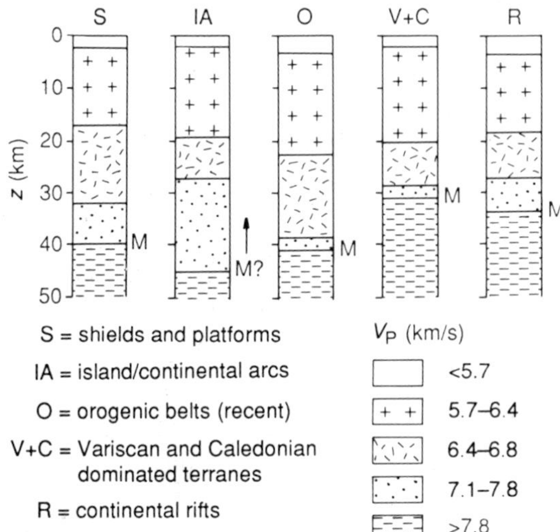

Figure 1.7 Structure and P wave velocities of various crustal types. (After Mooney and Meissner, 1991)

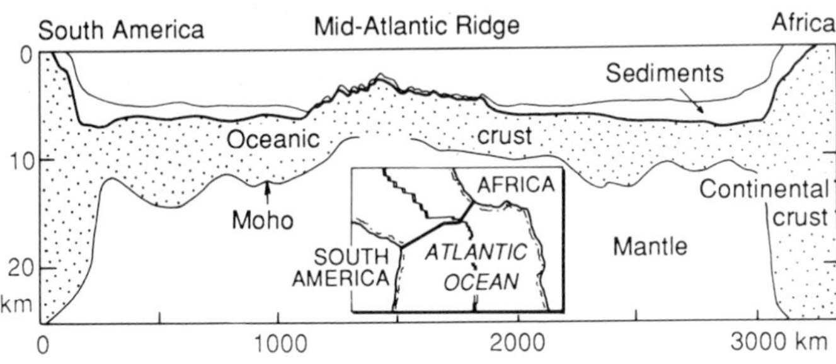

Figure 1.8 The general crustal structure of oceans and continental borders as shown by a cross-section of the Atlantic. (After Allegre, 1988)

and a light spot reflected off a dish of mercury. He was influenced by the writings of Michell. Knott wrote on the theory of reflection and refraction in 1899, and Wiechert (1910) published a paper describing the refraction method. The development of the seismic method for exploration was stimulated by sonic ranging systems and the recording of ground waves in World War I to locate guns by tracing air waves and the seismic waves from the reaction of the heavy guns on the ground, back to their source (Weatherby, 1940).

In seismic exploration the most common method is the *reflection* technique (see Chapter 15), in which many small seismometers (detectors) or geophones are laid out, usually in a line and often at comparatively short distances from the seismic sources (explosions or vibrators) so as to record near-vertical reflections from various layers (Figure 1.10). After a number of corrections and sophisticated data processing (Chapters 17–21) seismic sections like geographical cross-sections are produced (Figure 1.11).

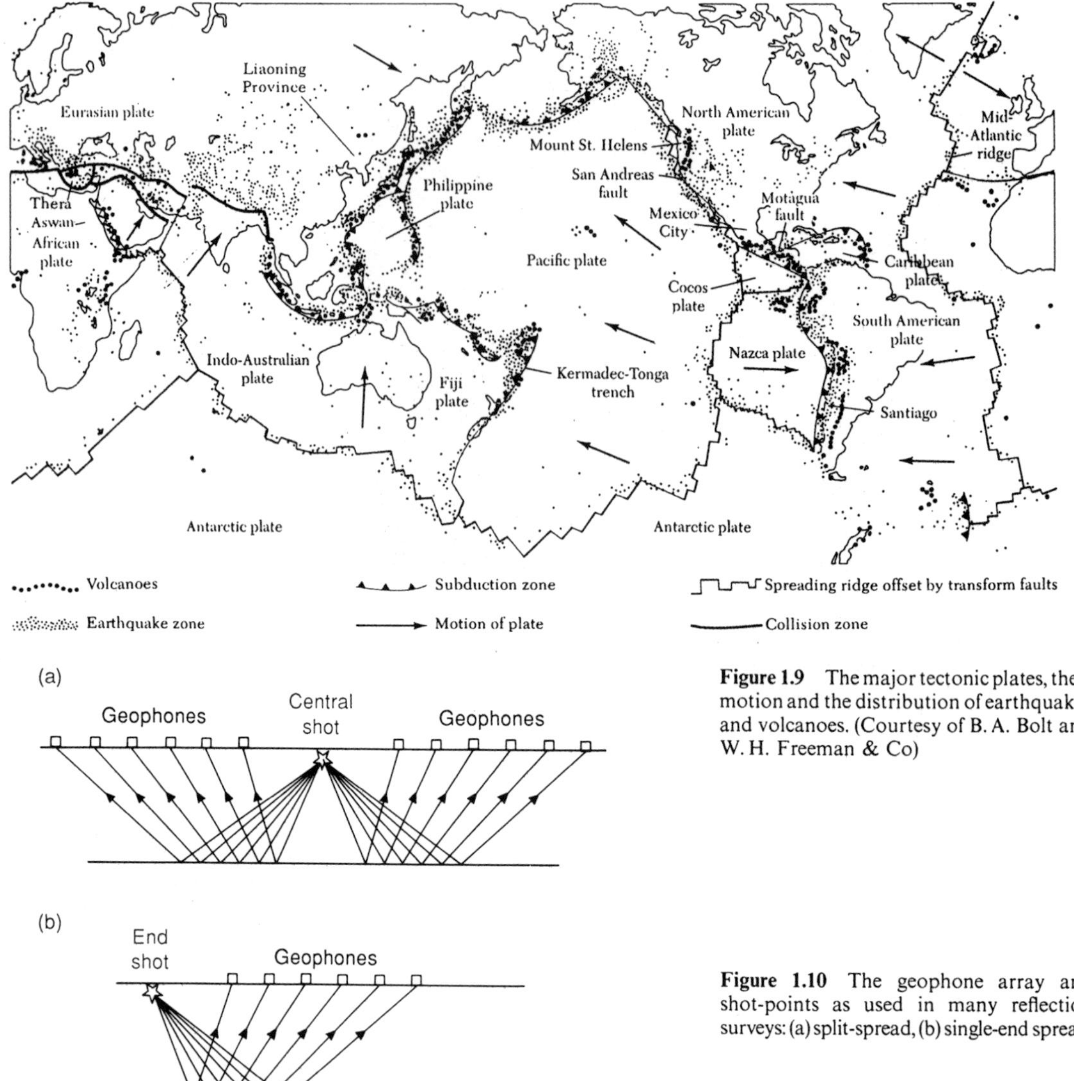

Figure 1.9 The major tectonic plates, their motion and the distribution of earthquakes and volcanoes. (Courtesy of B. A. Bolt and W. H. Freeman & Co)

Figure 1.10 The geophone array and shot-points as used in many reflection surveys: (a) split-spread, (b) single-end spread

The *refraction* method (Chapter 14) is designed to record waves refracted (bent) back to the surface because of the increasing velocity of layers with depth. The main part of the path is usually near-horizontal, along an interface, rather than vertical. The geophones are placed at larger distances, four or more times the expected depth to a refracting layer (Figure 1.12). This method is simpler than the reflection technique and gives average velocities and depths of refractors but does not usually image the sub-surface in sufficient detail for modern hydrocarbon exploration. The two methods are somewhat complimentary, reflections providing accurate structural information and refraction providing extra velocity data.

Figure 1.11 A block diagram of the results of a 3D reflection survey. Sections may be displayed in any direction through the block

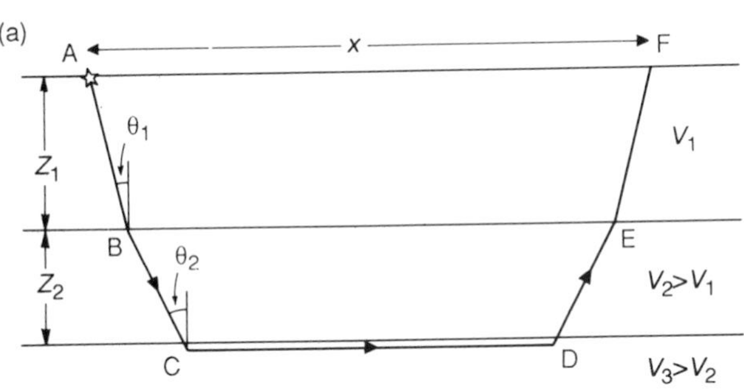

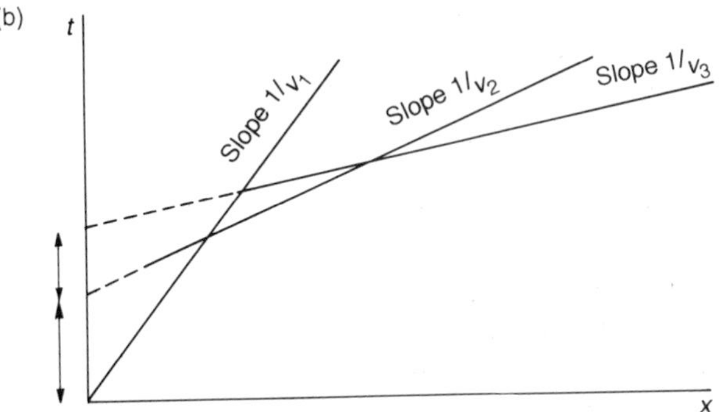

Figure 1.12 (a) Seismic refraction through a three-layer model. (b) Travel-time curves for the direct wave at velocity (V_1) and the critically refracted waves (V_2 and V_3) from two deeper layers. (After Kearey and Brooks, 1991)

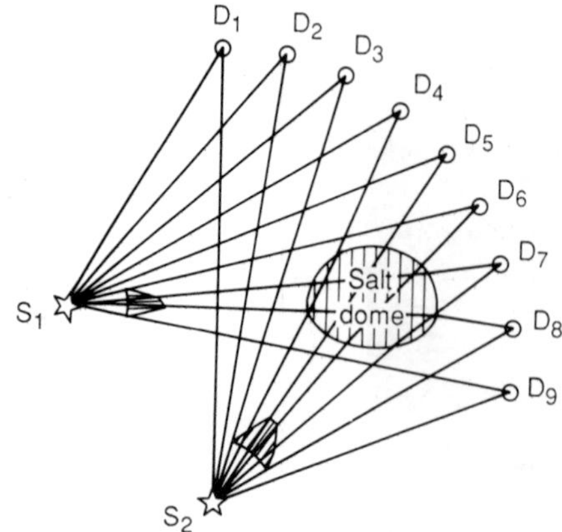

Figure 1.13 Fan shooting using direct or refracted waves to detect anomalous velocities. (After Kearey and Brooks, 1991)

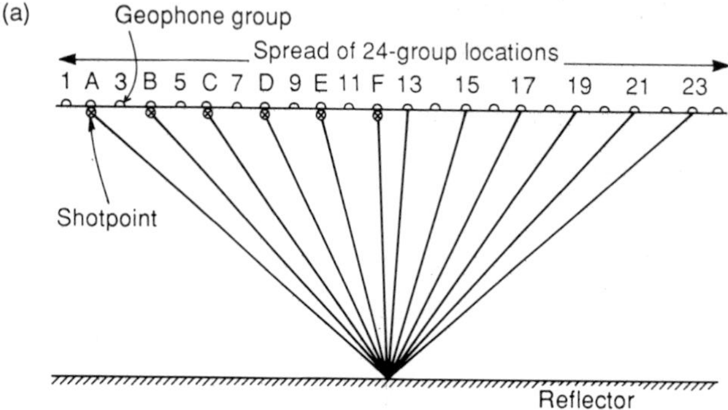

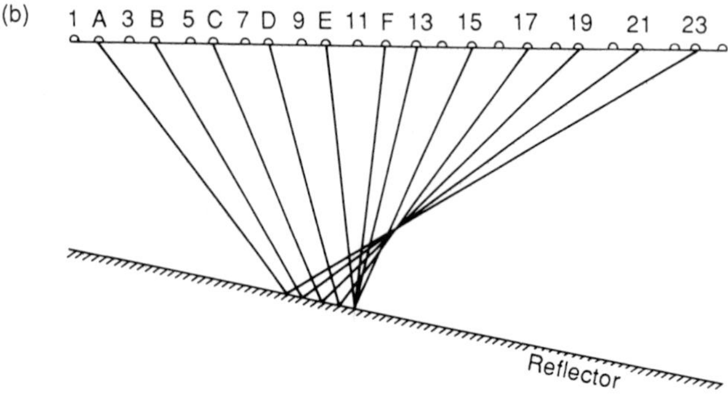

Figure 1.14 Common-mid-point shooting in reflection surveys. (a) Six-fold with 24 geophone groups. The shotpoint is moved two group intervals after each shot so that each reflection point (strictly small area) is sampled six times. (b) If the reflector dips, the data are smeared somewhat. (After Sheriff, 1984)

In 1919, Mintrop, Wiechert's student, applied for a patent for the refraction method to locate and measure the depths of sub-surface features. The earliest use was in Mexico and later the Texas Gulf Coast in 1923 (fan shooting, Figure 1.13). Mintrop also developed a mechanical-optical seismograph. The early mechanical seismographs were soon replaced by more sensitive electrical ones. Refraction (and gravity) surveys became successful in finding salt domes in Texas, e.g. the Orchard Dome, and later the large oil-bearing structures in Iran.

Meanwhile seismic reflection experiments were first attempted by Karcher in Oklahoma in 1919–1920, where he successfully mapped a shallow reflective dipping bed using a vacuum tube amplifier (Allen, 1980). By the early 1930s the reflection method had become the most widely used of geophysical methods with the need to map structure in more detail. Dips were calculated from differences in reflection travel-time. Electrical filtering of recordings was used to reduce noise, and radios were used for the explosion time (Macelwane, 1940). By the end of the 1930s, systems had up to 12 channels, automatic volume control and six geophones per group. Seismic surveys in the shallow bays of the Gulf coast began in the late 1920s but not in the open ocean until the late 1940s. By the 1960s geophysical surveys, including reflection and refraction, were being carried out on almost all the world's continental shelves. Explosive sources have been largely replaced by the Vibroseis method on land (developed in 1953) and the air-gun at sea in reflection surveys.

Analogue magnetic tape recording and playback became available around 1953, allowing greater flexibility, repeat shots, summing, etc. In the same year, the application of information theory to seismic data was begun by Enders Robinson and the MIT group (Clark, 1985), introducing deconvolution methods (p. 167). Common depth recording (or common mid-point, Figure 1.14), was patented in 1956 by Mayne and became widespread in the early 1960s (Mayne, 1967). Transistorized electronics came in 1959. An important breakthrough in the 1960s was that of digital recording and data processing, allowing computer applications, suppression of noise and multiple reflections, migration of data, and the determination of velocities and statics corrections (Hatton *et al.*, 1986; Yilmaz, 1987). In the 1980s interactive computer workstations arrived (Figure 21.1), as well as colour displays, much more detailed '3D' surveys, vertical seismic profiles and the use of S velocities in the reflection method. Some aspects of the development of exploration geophysics are treated in Bates *et al.* (1982).

Chapter 2
Seismic waves

2.1 SEISMIC WAVES, BODY WAVES

If the equilibrium of a solid body (such as the Earth) is disturbed, e.g. by a fault motion (earthquake) or explosion, two seismic (elastic) waves are transmitted through the body (Figures 1.2 and 1.3). The waves are often assumed to be radiating from a point (the *focus*), although sources are extended in earthquake faulting. The point on the surface above a focus is the *epicentre*.

The first waves to arrive are the P ('pressure', compressional) waves, ordinary sound waves, also called longitudinal waves because ground motion is a vibration in the direction of transmission (Figure 2.1). The name P wave comes from the Latin *primus* ('first') for the first waves to arrive, having the highest velocity. The second waves to arrive are the S (*secundus*) waves which have a transverse, shear vibration in a plane perpendicular to the direction of propagation (Figure 2.1).

The reason there are two types of waves and elastic constants in a solid body is that there are two fundamental ways that you can strain a solid body: (1) by volume change without change of shape, i.e. pure compression or expansion (Figure 2.2); or (2) by change of shape without change of volume, i.e. a distortion.

It is believed that the Roman architect Vitruvius first suggested the analogy between sound propagation and wave motion, as on water, about 20 BC. Much later, Newton first estimated the velocity of sound in air using Hooke's Law for springs and Boyle's ideas on the elasticity of the air.

Box 2.1 Elastic strain and the elastic constants

As a P wave passes through a body, *strain* is produced as defined by

$$s = \frac{\delta u}{\delta x}$$

i.e. it is the ratio of the change in distance between two points in the body to the original distance. These are linear strains, considering the straining in one direction. Likewise, in the other two directions, y and z, the strain would be given by

continued

continued

$$\frac{\delta v}{\delta y} \quad \text{and} \quad \frac{\delta w}{\delta z}$$

Similarly we have volume strain, or dilation, $\mathrm{d}V/V$ (δV is the change in volume) and shear strain ψ (Figure 2.2).

Now for some definitions for the elastic constants in a solid. In geophysics the elastic constants commonly used are the incompressibility or bulk modulus k and the rigidity μ. Mathematicians tend to use the Lamé constants λ and μ, while engineers use Young's modulus Y and Poisson's ratio σ. Young's modulus is for cases of simple extension or contraction δl (linear strain).

$$\text{Incompressibility or Bulk Modulus } k = \frac{\text{stress}}{\text{strain}} = \frac{\partial P}{\partial V/V} = \frac{\partial F/A}{\partial V/V}$$

(Figure 2.2)

where $\frac{\partial V}{V}$ is the proportional change in volume (called dilation) under hydrostatic pressure.

$$\text{Shear Modulus or Rigidity } \mu = \frac{\text{stress}}{\text{strain}} = \frac{\partial P}{\psi} = \frac{\partial F/A}{\partial x/l}$$

(Figure 2.2)

$$\text{Young's Modulus } Y = \frac{\text{stress}}{\text{strain}} = \frac{\partial P}{\partial l/l}$$

$$\text{Poisson's Ratio } \sigma = \frac{\text{transverse strain}}{\text{longitudinal strain}} = \frac{\partial x}{\partial l}$$

When an elastic body is stretched, for example, it is lengthened and thinned; if compressed, it is shortened and widened, by ∂l and ∂x, respectively. σ is also given by

$$\frac{V_p^2 - 2V_s^2}{2(V_p^2 - V_s^2)}$$

and is usually 0.25 for solids, where V_p and V_s are the P and S wave velocities.

So we have P waves, or compressional waves, which transmit pressure changes through the Earth as a series of alternating compressive and rarefaction (decompressive) waves. They may be audible close to an earthquake epicentre. Being the first to arrive, P waves are the most accurately measured and are the waves most commonly used in earthquake location and in the reflection and refraction methods of exploration. Their velocity is given by

$$V_p = \frac{(k + 4/3\mu)^{1/2}}{(\rho)}$$

where k is the incompressibility (the inverse of the compressibility; see Box 2.1) or bulk modulus of the rock, μ is the rigidity, and ρ the density (see Box 2.2 for a derivation). It is worth noting that the strains produced in the rock as seismic waves pass through are normally

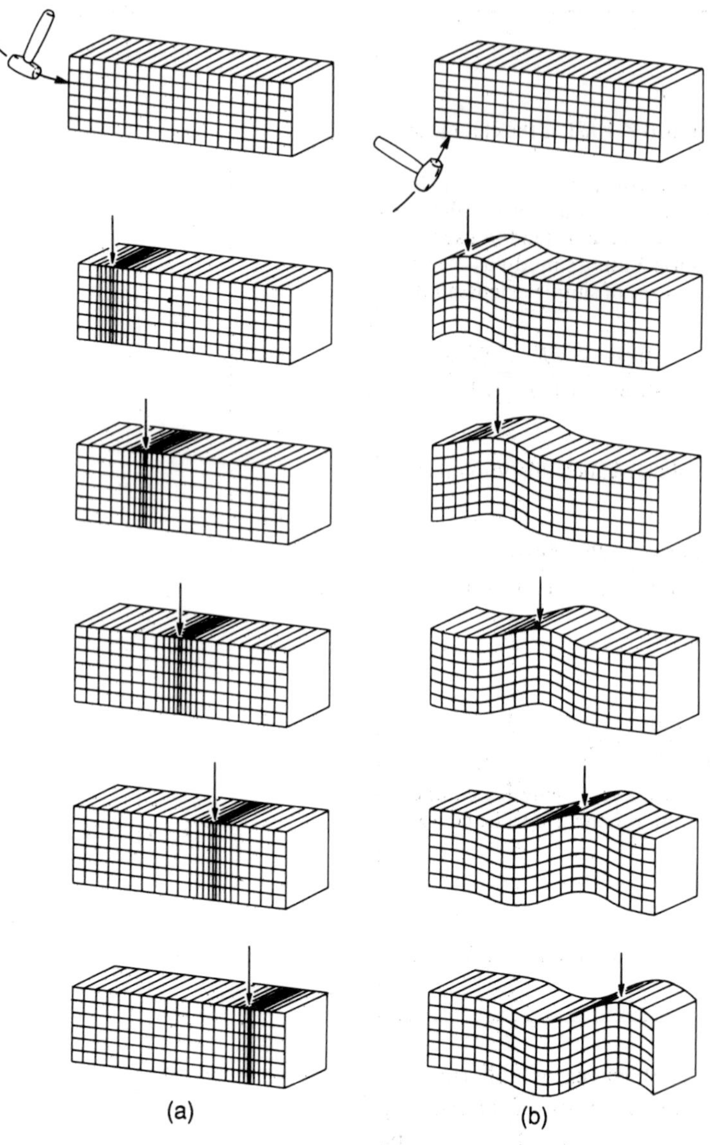

Figure 2.1 The body waves P and S. (a) P waves cause longitudinal ground motion, along the direction of travel; (b) S waves cause transverse vibration. (After Uyeda, 1978)

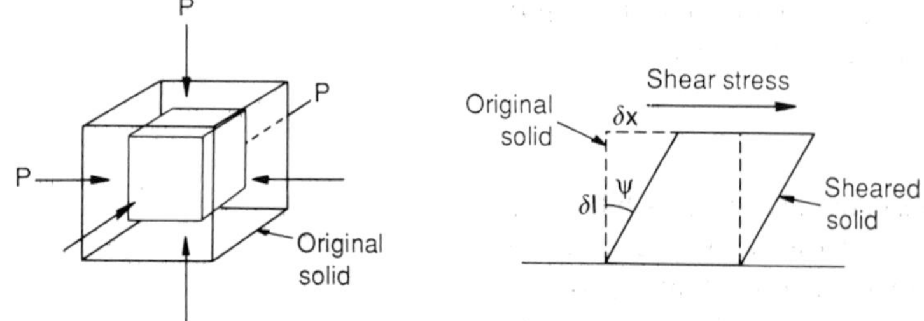

Figure 2.2 (a) Deformation by change of volume due to compression; (b) shear strain for case of block with lower side fixed. (After Burger, 1992)

extremely small, about one part in a million (10^{-6}), except very near a source, so that linear (proportional) relations between stress and strain can be assumed.

The S waves, the shear or transverse waves, have a velocity given by

$$V_s = (\mu/\rho)^{1/2}$$

This velocity is about 60% of the P velocity and since it arrives within some noise from the P arrival, it is more difficult to time accurately. However, S waves are valuable in giving information on rigidity, and possible fluid content because P waves, but not S waves, are transmitted through fluids, as fluids have no shear strength (rigidity $\mu = 0$). This is one reason why we know that the outer core of the Earth is fluid, almost certainly mostly liquid iron, as has been deduced from geochemical and magnetic data (see Chapter 13). An S_v wave is one in which the ground vibration is vertical and an S_h wave is one in which it is horizontal.

P and S waves travel through the body or interior of the Earth, in distinction from surface waves, and so are called body waves (Figure 1.2). Thus seismic velocities are governed by the elastic constants and density of the rocks in the interior, particularly the former as the range of densities in the Earth is smaller. In turn these values depend on rock composition, pressure and temperature. As the depth in the Earth increases, the elastic constants increase more than does the density, resulting in a general increase in velocities with depth and a curvature of the ray paths towards the surface (Figure 1.2).

Box 2.2 Derivations of the velocities of P and S

A derivation of the velocity for P is given, partly following Stacey (1977). More detailed and rigorous analyses can be found in various texts such as Aki and Richards (1980), Bullen and Bolt (1985) and Dobrin and Savit (1988).

First, for the simpler case of a P wave in a long bar unconstrained at the sides, the only elastic constant involved is Young's modulus (Y) which, as we saw earlier, is a measure of the elasticity of extension and contraction. Consider a small element of a bar with cross-section A. The compressional stress (force per unit area, like a pressure) across the face is S, and at a point dx further along the bar $S + (\delta S/\delta x)\delta x$, i.e. there is a gradient in the stress. If the displacement of the element due to the passage of the wave is μ, the mass × acceleration of the element is equal to the difference in force at the two ends, i.e. $\rho A\, \delta x(\delta^2\mu/\delta t^2) = A(\delta S/\delta x)\,\delta x$, where ρ is the density, thus

$$\rho(\delta^2\mu/\delta t^2) = \delta S/\delta x$$

Young's modulus Y is defined as the longitudinal stress divided by the longitudinal strain, i.e. Y = stress/strain = $S/(\delta\mu/dx)$, as the strain is the differential displacement by definition, the ratio of the change in length over a distance δx, so

$$S = Y\delta\mu/\delta x \text{ and } \delta S/\delta x = Y(\delta^2\mu/\delta x^2)$$

and

$$Y(\delta^2\mu/\delta x^2) = \rho(\delta^2\mu/\delta t^2)$$

continued

continued

or

$$(\delta^2\mu/\delta t^2) = Y/\rho(\delta^2\mu/\delta x^2)$$

This is a wave equation and it is easy to show by differentiating that a travelling wave $\mu = f(V_p - x)$ satisfies it with the velocity $V_p = \sqrt{(Y/\rho)}$. This is the velocity of a P wave in a long rod, i.e. where the wavelength (the distance between wave peaks) is much bigger than the width of the bar.

P and S waves are sound and elastic waves in the Earth and so their velocities also depend on the ratios of the appropriate restoring elastic forces to the inertias (densities) of the rock.

Since S is a shear wave, μ the rigidity is the correct elastic modulus for S and so $V_s = \sqrt{(\mu/\rho)}$, but that for P is less obvious. In the passage of P waves, the Earth undergoes a succession of compressions and rarefactions (decompressions), but with lateral constraints of the surrounding rock allowing only longitudinal vibration, unlike the case of a rod transmitting compressional waves where lateral movement and strain is possible.

With no lateral restraint a longitudinal stress S_1 would produce a longitudinal strain L related through Young's modulus (Y) as we saw. Thus $L_1 = (S_1)/Y$, and two perpendicular lateral strains would also be produced given by Poisson's ratio (σ), thus $L_2 = L_3 = -\sigma L_1 = -\sigma S_1/Y$.

Now superimposing the strains for all three stresses we get

$$L_1 = 1/Y[S_1 - \sigma(S_2 + S_3)]$$
$$L_2 = 1/Y[S_2 - \sigma(S_3 + S_1)]$$
$$L_3 = 1/Y[S_3 - \sigma(S_1 + S_2)]$$

However, in a constrained medium, such as inside the Earth, the stresses S_1 and S_2 adjust to make zero the strains perpendicular to the wave direction, thus

$$S_2 = \sigma(S_1 + S_3) \qquad \text{and} \qquad S_3 = \sigma(S_1 + S_2)$$

so

$$S_2 + S_3 = (2\sigma S_1)/(1 - \sigma) \text{ and } L_1 = (S_1/Y)[1 - (2\sigma^2)/(1 - \sigma)]$$

and the appropriate elastic modulus for the longitudinal P wave is

$$m = S_1/L_1 = Y(1 - \sigma)/(1 - 2\sigma)(1 + \sigma)$$

There are known relations between the elastic constants which will allow us to convert this equation to one in the elastic constants k and μ instead of using Y and σ.

$$k = Y/[3(1 - 2\sigma)] \qquad \text{and} \qquad \mu = Y/2(1 + \sigma),$$

hence

$$m = k + 4/3\mu$$

and

$$V_p = \sqrt{[(k + 4/3\mu)/\rho]}$$

and as we saw

$$V_s = \sqrt{(\mu/\rho)}$$

Since k is normally about 1.7μ, V_p is commonly $1.7V_s$ and V_s about $0.6V_p$.

Velocities in various rock types have been determined in drilled cores and other samples in the laboratory (Christensen and Wefner, 1989). P-wave velocities vary from near the velocity of sound in air and in aerated soil, i.e. about 0.3 km/s, to 1.45 km/s in the ocean, 3–5 km/s in sediments, 5–6 km/s in basement rocks and the upper crust, to 8 km/s in the upper mantle, over 13 km/s in the lower mantle, and 11 km/s at the centre of the Earth (see Table 2.1 for velocities in sediments; also Clark, 1966; Christensen, 1989). P waves travel right through to the opposite side of the Earth in 18 min.

Jeffreys and Bullen (1940) produced the 'J-B' tables for the travel-times of the various phases through the Earth to all distances. The tables have been a standard used for many years in the location of

Table 2.1 Compressional-wave velocities in Earth materials

	V_p (km/s)
Unconsolidated materials	
Sand (dry)	0.2–1.0
Sand (water saturated)	1.5–2.0
Clay	1.0–2.5
Glacial till (water saturated)	1.5–2.5
Permafrost	3.5–4.0
Sedimentary rocks	
Sandstones	2.0–6.0
Tertiary sandstone	2.0–2.5
Pennant sandstone (Carboniferous)	4.0–4.5
Cambrian quartzite	5.5–6.0
Limestones	2.0–6.0
Cretaceous chalk	2.0–2.5
Jurassic oolites and bioclastic limestones	3.0–4.0
Carboniferous limestone	5.0–5.5
Dolomites	2.5–6.5
Salt	4.5–5.0
Anhydride	4.5–6.5
Gypsum	2.0–3.5
Igneous/metamorphic rocks	
Granite	5.5–6.0
Gabbro	6.5–7.0
Ultramafic rocks	7.5–8.5
Serpentinite	5.5–6.5
Pore fluids	
Air	0.3
Water	1.4–1.5
Ice	3.4
Petroleum	1.3–1.4
Other materials	
Steel	6.1
Iron	5.8
Aluminium	6.6
Concrete	3.6

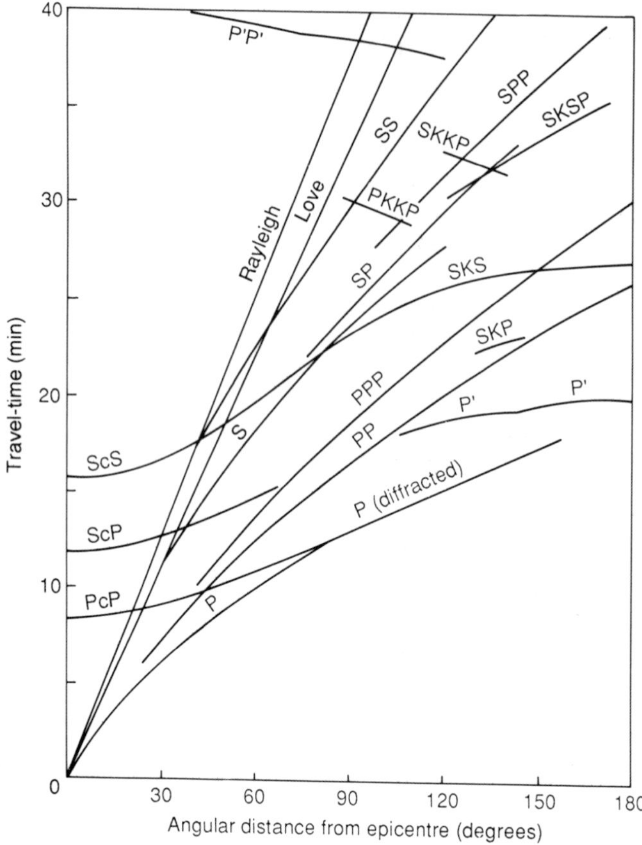

Figure 2.3 Travel-time curves for waves from a shallow focus. (After Jeffreys and Bullen, 1940)

earthquakes. They have recently been replaced by the work of a sub-commission of the International Association of Seismology and the Physics of the Earth's Interior (Kennett and Engdahl, 1991). Such tables include the travel-times of reflections from the surface such as PP and SS (Figure 1.2) and from the core such as PcP and ScS, as well as those converting between P and S, such as PS and SP (Figure 2.3 and Tables 2.2 and 2.3). The new tables include the corrections to the main phases for the ellipticity of the Earth.

2.2 SURFACE WAVES

Just as there are two main kinds of body waves (P and S) which travel through the interior, there are two types of surface wave, the Rayleigh and Love waves which travel near the surface of the Earth (Figures 1.3 and 2.4). Thus there are two classes each of body waves, surface waves and elastic constants. Rayleigh (1885) predicted the existence of surface waves when he mathematically modelled the motion of plane waves in an elastic half-space. Love (1911) investigated the effect of a

Table 2.2 A short travel-time table

Distance (degrees)	Surface focus P min	s	S min	s	S–P min	s
10	2	28	4	22	1	54
20	4	37	8	17	3	40
30	6	13	11	10	4	57
40	7	38	13	45	6	07
50	8	58	16	09	7	11
60	10	11	18	23	8	12
70	11	15	20	26	9	11
80	12	13	22	17	10	04
90	13	03	23	55	10	52
100	13	48	25	20	11	32

Table 2.3 S–P travel-time differences for various distances and focal depths

Distance (degrees)	Depth (km) 100		300		500		700	
20	3	34	3	24	3	12	3	07
30	4	51	4	37	4	26	4	16
40	5	57	5	41	5	27	5	17
50	6	59	6	44	6	31	6	19
60	8	02	7	44	7	30	7	18
70	8	54	8	37	8	22	8	09
80	9	48	9	31	9	18	9	05
90	10	42	10	25	10	10	9	56
100	11	26	11	09	10	51	10	37

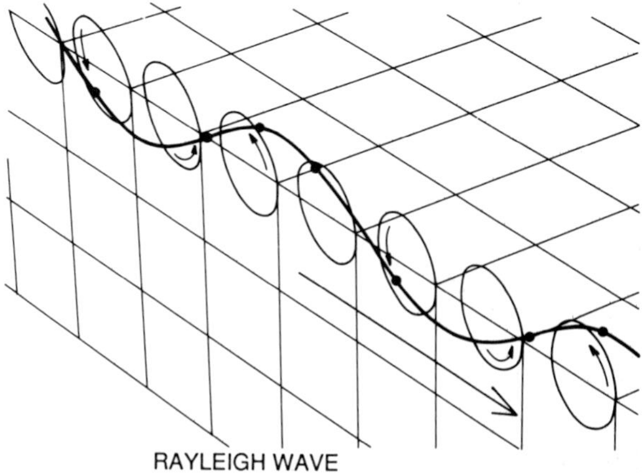

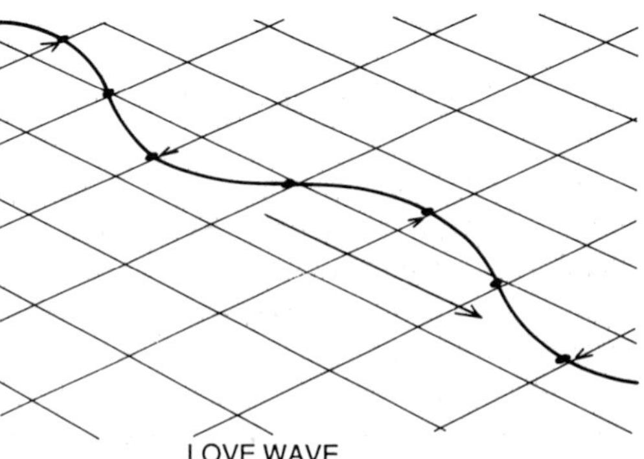

Figure 2.4 Surface waves; Love waves produce transverse waves in the horizontal plane, while Rayleigh waves cause a rolling motion like ocean waves with an elliptical ground motion in the vertical plane, but backwards at the top of the ellipse, see Figure 2.5. (After Oliver, 1959)

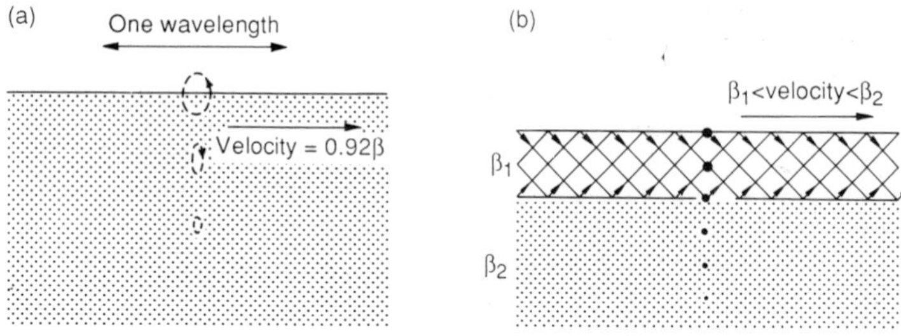

Figure 2.5 (a) Rayleigh-wave ground motion with depth in a uniform solid where β is the S velocity. (b) Love waves as the result of interference between waves reflected many times in the upper layer. Ground motion is horizontal in Love waves. (After Bott, 1982)

surface layer and discovered the other important surface wave, which was named after him.

If a disturbance, such as an earthquake or explosion, occurs near the surface (within a wavelength or so), a significant amount of the seismic energy is trapped near the surface and is transmitted as surface waves which travel as 'guided' waves. Their amplitudes decrease with depth (Figure 2.5). Surface waves may be regarded as the result of the addition and interference of waves incident and reflected at the surface of the Earth (plus diffraction effects when the curved wave-fronts from the focus reach the surface). The surface waves are trapped not only by the ground–air interface, which is a good reflector, but also by the crust and low-velocity layers near the surface.

So there are two types of surface wave: (1) Rayleigh waves in which the ground motion is elliptical, as for a water wave (but retrograde, Figure 2.5); and (2) the faster Love waves, which are shear waves like S waves, but vibrate only in the horizontal plane (horizontally polarized). Surface waves travel in two dimensions, not three as for body waves, as they are guided by the surface, so their amplitudes decrease more slowly with distance and they tend to dominate recordings at a large distance from shallow earthquakes.

The velocities of the main types of seismic waves are, in descending order, P velocity, S velocity, Love and Rayleigh waves, i.e. $V_p > V_s > V_{Love} > V_{Rayleigh}$; whereas amplitudes are generally in the reverse order, with Rayleigh waves having the largest amplitude and P the smallest. The amplitudes of surface waves decrease with the depth of the focus and so the surface wave amplitudes relative to the body waves are an approximate indication of the depth of the source. A special type of Rayleigh wave (named after Stoneley) occurs at interfaces between layers inside the Earth.

Love waves require the presence of a surface layer of lower velocity, such as the crust, or an increase of velocity with depth, i.e. a velocity gradient. They would not occur on a uniform plane solid as do Rayleigh waves and can be understood as the summation of multiple reflections in, for example, a simple one-layered crust (Figure 2.5b). Love waves in a layer have a maximum velocity of $0.9V_s$ and in a simple one-layered model for the Earth (which is useful for shorter distances) Love wave velocities have the upper and lower medium S velocities as extremes. Note that the ocean, having no rigidity, does not affect Love waves but does influence Rayleigh wave dispersion.

Seismograms (Figure 1.3) show that surface waves are not short wavelets as for P and S waves, but spread out in time and space. This 'dispersion' of the surface waves is caused by their velocities being a function of wave period (the time between peaks and the inverse of frequency) and of wavelength. The longer period surface waves (and thus longer wavelength) are affected by deeper material in the Earth where elasticities, velocities and densities (to which they are also sensitive) are generally higher. For example, Rayleigh waves are very sensitive to the rigidity at depths of about 0.4 of the wavelength. Because of dispersion the longer period waves arrive first and the shorter periods are strung out later in the record (Figure 1.3). It causes an initial pulse to broaden out into a train of waves as it travels further. Dispersion can also be seen in ocean waves and in ripples on water (their restoring force is gravity and surface tension respectively, not elasticity). In a homogeneous half-space there would be no such dispersion, and no Love waves. Additional dispersion of seismic surface waves is produced by the Earth's curvature. The later part of the surface wave train, the coda, is related to scattering by velocity inhomogeneities.

Box 2.3 Phase and group velocities

Surface-wave velocities are measured either as phase velocities (the velocity of some peak, trough or wave of single frequency), or as the velocity of a group of waves (Figure 2.6) which is also the velocity of energy transmission. Group velocity $U = V - \lambda \mathrm{d}V/\mathrm{d}\lambda$, where V is the phase velocity and λ the wavelength. The group velocities can be simply measured from the arrival time of groups from an epicentre of known position and time. However, the phase velocities are measured by the travel-time of a peak or trough between two long-period stations along a path from the epicentre or time across an array of stations.

On short-period records of local or regional earthquakes the short-period surface wave Lg usually dominates. This is interpreted as the result of multiply reflected waves in the crust or the result of the superposition of higher mode surface waves. For periods of less than

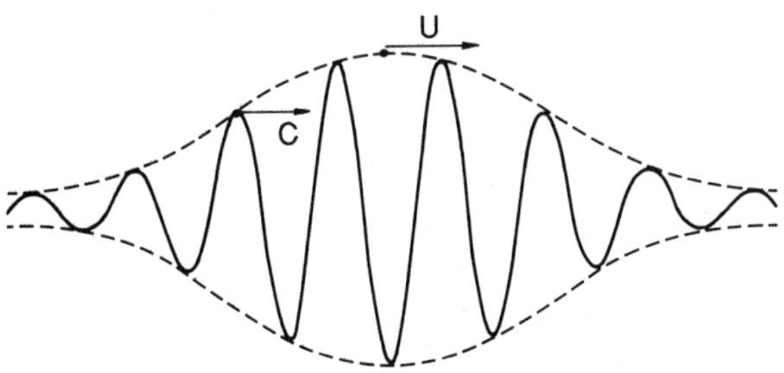

Figure 2.6 Surface waves have velocities which vary with period and wavelength, so producing groups of waves. The phase velocity (C) is the velocity of an individual peak or trough, the group velocity (U) that of a group of waves. (After Bott, 1982)

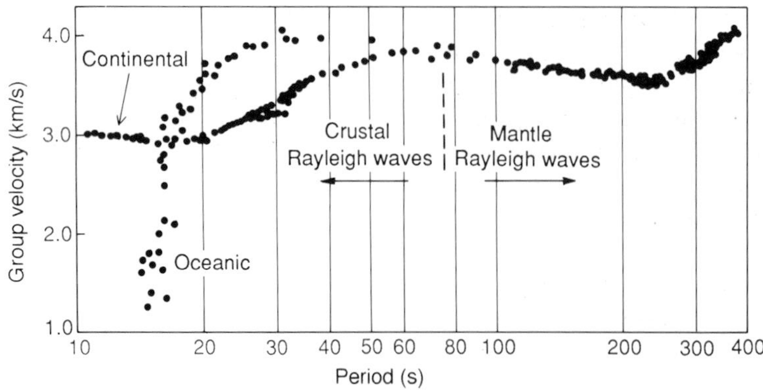

Figure 2.7 The variation of group velocity of Rayleigh surface waves with period (dispersion). (After Ewing and Press, 1950)

about 15 s sedimentary layers affect surface wave velocities strongly. For periods of about 15–300 s the crust and upper mantle structures most strongly affect dispersion (Figure 2.7). Gutenberg (1924) first suggested that oceanic crusts were thinner than continental crusts when dispersion studies showed faster surface waves over the oceans. Evidence for the low-velocity region in the mantle was seen in the new very long period data obtained in the 1960s by Dorman *et al.* (1960). If models of the Earth are set up, theoretical dispersion curves can be calculated and compared with observed data (Kovach, 1978). Simpler flat Earth models can be used for Rayleigh waves of period less than about 60 s, but for Love waves only for less than 15 s. The accumulation of a large database of long-period digital recordings from the Global Digital Seismic (GDSN) and accelerometer networks in the 1980s has allowed analysis of surface-wave dispersion on a global scale.

A complementary approach is to use the mathematical techniques called inversion which take the observed data, plus some physical assumptions and limits, and compute a range of possible models (e.g. see Woodhouse and Dziewonski, 1984). Thus dispersion data may be inverted (if good enough) to provide information about the crust and mantle, and even whole Earth structure, particularly by including Earth oscillation periods. Our models of the Earth are thus produced by a combination of studies of body-wave travel-times and amplitudes, surface-wave dispersion, Earth oscillation periods, and recently by tomographic methods (see p. 97). Synthetic seismograms can also be computed (e.g. Chapman, 1978).

2.3 VELOCITY ANISOTROPY

It has been recognized in recent years that seismic anisotropy may be significant in most rock types of the crust and upper mantle, e.g. S waves in the crust may travel at different speeds depending on their polarization (vibration direction), and P waves on the direction of travel (Figures 2.8 and 2.9). Thus shear waves polarized in one

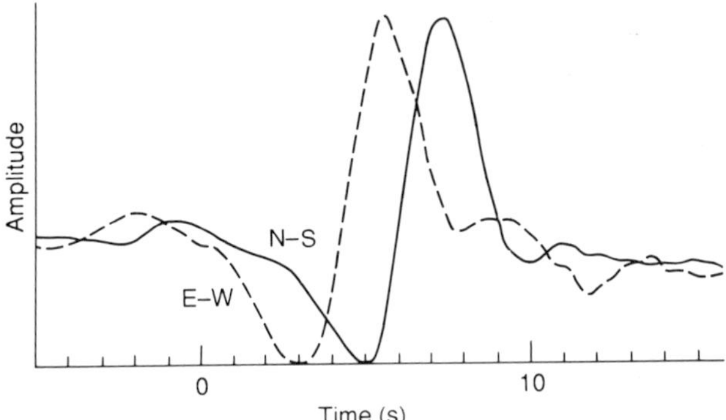

Figure 2.8 Shear-wave splitting. The north–south component of ground motion (solid curve) arrived 2 s after the east–west component (dashed curve). (From Ann. Report of the Terrestrial Magnetism Section of the Carnegie Institute of Washington, 1990, with permission)

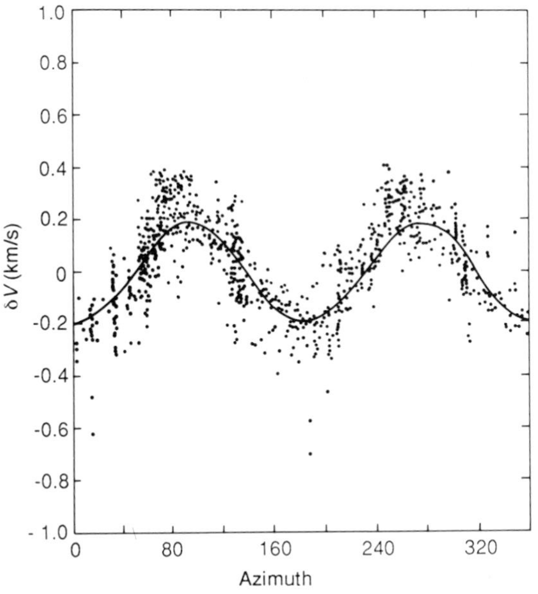

Figure 2.9 Azimuthal anisotropy of Pn waves in the Pacific upper mantle with a mean velocity of 8.16 km/s (Morris *et al*., 1969). The curve is the velocity measured in the laboratory for ophiolite samples from the Bay of Islands. (After Anderson, 1989a)

direction travel faster than those polarized at right angles to that direction. This 'shear wave splitting' is interpreted in terms of aligned and fluid-filled fractures consistent with stress directions (Crampin, 1984; Leary *et al*., 1990; Crampin and Lovell, 1991). Other causes of anisotropy may be layering and the preferred orientation of minerals. Some P waves in the upper mantle show anisotropy, e.g. with the presence of the mineral olivine (Dziewonski and Anderson, 1983). Such anisotropy can be valuable in providing information on rock fabric (Fountain and Christensen, 1989) and on former mantle flow directions (Figure 2.10).

The two special cases of anisotropy are (a) where speeds in a horizontal direction are different from those in a vertical direction, and (b) azimuthal anisotropy in which horizontally travelling waves

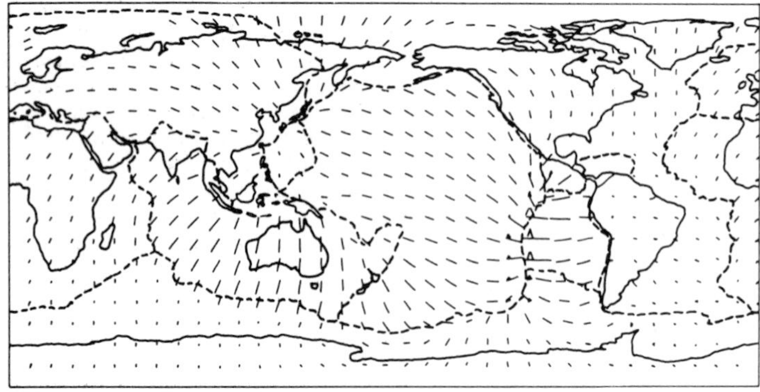

Figure 2.10 Flow lines at 260 km depth for a model of large-scale flow in the upper mantle. (After Hager and O'Connell, 1979)

have speeds differing with direction. In the former case the two polarizations of shear waves, SV vertical and SH horizontal, would have different speeds (Dziewonski and Anderson, 1984). This would also affect surface waves. Shear-wave splitting determined from surface-wave studies has shown fast directions correlating with the directions of convective flow of the material in the upper mantle below the lithosphere or plates, including vertical flow beneath the ridges and subduction zones (Figure 2.10) (Nataf *et al.*, 1986). Since the dispersion of the vertically oscillating Rayleigh waves depends largely on SV velocities, and the horizontally oscillating Love waves are affected mainly by SH velocity, surface waves are used to map such anisotropic differences (Dziewonski and Anderson, 1984). For example, Love waves are relatively fast in regions of horizontal mantle convective flow, related to plate motion, and slow in areas of ascending and descending flow. Rayleigh waves are slow over tectonic and young oceanic areas, but fast over shields and older oceanic areas (Figure 2.7).

2.4 OSCILLATIONS OF THE EARTH

Large surface waves and body waves travel around the Earth many times in all directions from the focus, they may thus interfere to produce standing waves and cause the Earth to vibrate as a whole, like a ringing bell (Lapwood and Usami, 1981). For example, Rayleigh waves travelling around the surface with an integral number of certain wavelengths can form standing waves, as on a vibrating string (Figure 2.11).

This effect is called 'oscillations of the Earth' and was apparently first observed by Benioff in 1952 following his development of the strain seismograph (p. 40 and Figure 2.12), and then in 1960 from the great Chilean earthquake by others using the newly developed long-period seismographs. Only earthquakes of about magnitude 6 or more can produce detectable Earth oscillations, but very large

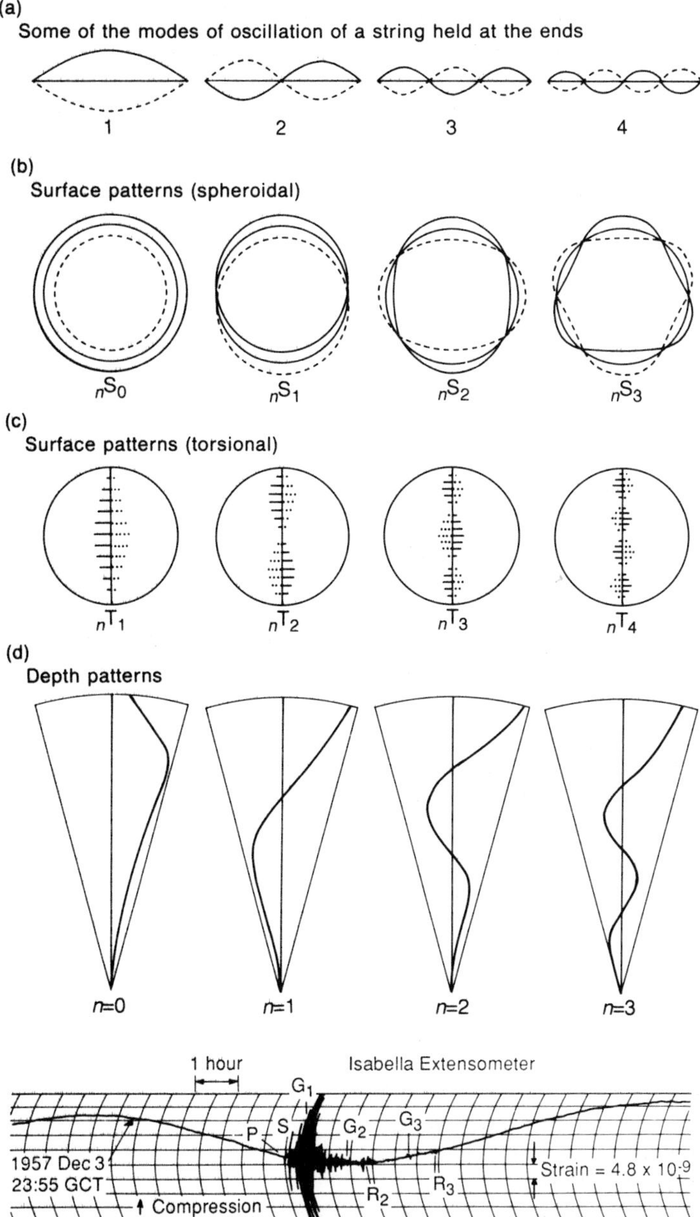

Figure 2.11 Modes of oscillation of a string and of a sphere. For the spheroidal (S) modes there is radial and horizontal vibration. In torsional (T) modes the vibration is on spherical surfaces only (horizontal). The subscripts 0–4 denote the variations in modes with latitude. In these examples the vibrations are assumed independent of longitude. The subscript n denotes the mode variation in depth. (After Press, 1965)

Figure 2.12 Record of the Mongolian earthquake of 1957 on a quartz strain seismograph at Isabella, California. G and R refer to Love and Rayleigh waves circling the Earth several times. Note also the slow effect of the Earth tidal strain. (After Benioff, 1959)

earthquakes can produce oscillations detectable for many days, even for up to a month afterwards for lower mode (slower) oscillations. The vibrations of an elastic sphere were first investigated mathematically by Poisson in 1829, and Lamb (1904) showed that there are again two classes of vibration. The theory was extended to non-homogeneous spheres by Alterman *et al.* (1959). Oscillations of the Earth are very useful for examining average Earth properties.

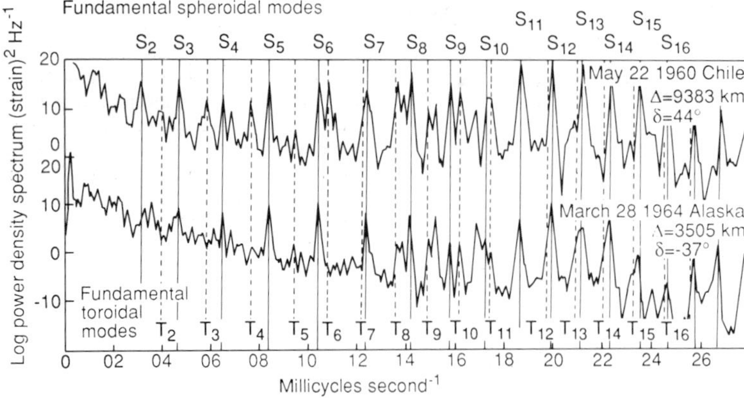

Figure 2.13 Some of the numerous modes of earth oscillation produced by the Chilean earthquake of 1960 and the Alaskan earthquake of 1964. (After Bott, 1982)

It should be noted that there are two ways to view seismic records, either as travelling waves, plotted against time (in the 'time domain'), or as global oscillations, a spectrum in the 'frequency or period domain'. Any earthquake vibration can be considered as the sum of many global vibration modes. So we have 'seismic spectroscopy' (Figure 2.13). Even a P or other body wave may be thought of as a summation of certain higher modes, too many usually to be practical. For periods less than 300 s wave types are easily recognized on a record, but for periods above 1000 s, wavelengths are about Earth radius in size and travelling waves are difficult to distinguish. They are then best studied as oscillations of the whole Earth. On the other hand, at high frequencies the large number of modes makes this approach difficult.

Figure 2.11 illustrates the various modes of oscillation. The lower modes of oscillations of the Earth have very long periods of many minutes, the longest being 54 min (the fundamental spheroidal or 'football' mode). Just as there are two types of body and surface waves, there are two modes of Earth oscillation, spheroidal and torsional (or toroidal), corresponding to Rayleigh and Love waves respectively. Spheroidal oscillations include pulsating modes (lowest frequency 20 min) and 'football'-shaped modes, while torsional modes are the various twisting oscillations of a sphere. The latter depend only on shear velocities and densities, whereas the spheroidal modes depend on both P and S velocities and density. Spheroidal oscillations can be detected by recording gravimeters, but not torsional oscillations because they produce no vertical ground movement.

About 1500 periods of Earth oscillations have been identified. As noted above, the comparison of observed periods with those calculated from Earth models has provided another method of examining the average structure of our planet. The oscillation spectra also show a splitting in peaks due to the Earth's rotation, and to ellipticity and regional differences in the Earth, e.g. the football mode is split into two peaks at 53.1 and 54.7 min (Masters *et al.*, 1982). Rotation of the

Earth affects the periods because the centrifugal effect is greater for waves travelling east (the same direction as the Earth rotates) than for waves travelling west, a change in effective gravity on the moving wave.

The Sun and stars are also believed to oscillate continually and the Sun's oscillation periods are being measured to assist study of its structure (Christensen-Dalsgaard *et al.*, 1985; Libbrecht and Woodward, 1991), e.g. estimation of the depth of the convective zone.

2.5 AMPLITUDES

The amplitude of the ground vibrations from a large earthquake hundreds of kilometres away may be as much as several centimetres for the surface waves. Surface waves from a distant focus, however, have such long wavelengths (hundreds of kilometres) and slow periods that they are not felt by people. Early seismographs, though of very low magnification, could record useful data from large shocks. A small local earthquake may be felt even though it causes ground motion of only a few millimetres; this is because of the higher wave frequencies (say 50 Hz to 0.1 Hz), sometimes high enough to be audible. Modern seismographs have high amplification (perhaps 100 000 or more; Figure 2.14) and can detect ground motion as small as several nanometres (10^{-9} m) in good conditions and record earthquakes from the other side of the Earth of magnitude 6 or less.

The amplitudes of reflected waves depend on the change in the seismic or acoustic impedance (ρv) across an interface. This product of the velocity and density is analogous to electrical impedance in that no change in impedance gives maximum transmission, but strong reflection requires a significant change in seismic impedance. The reflection coefficient R, the ratio of reflection amplitude A_2 to incident amplitude A_1, the normal (90°) incidence is given by

$$R = \frac{A_2}{A_1} = \frac{\rho_2/v_2 - \rho_1/v_1}{\rho_2/v_2 + \rho_1 v_1}$$

where v_2,ρ_2 and v_1,ρ_1 are the velocities and densities in the lower and upper layers respectively. For other angles of incidence the Zoeppritz equations apply (Figure 2.15) (Appendix 2). Shuey (1985) simplified the equations.

Reflection coefficients between different rock types are commonly less than 0.1, although the base of the weathered layer could have a value of about 0.4. Note that reflections could occur without a change in velocity but with a change in density, although in practice velocity changes are much greater than for density. Also, the reflection coefficient R will be negative if the product ρv reduces across the boundary (usually because of a velocity decrease). A negative value means a reversal in the phase of the reflection, a negative amplitude. This occurs at the ocean surface where the velocity and density

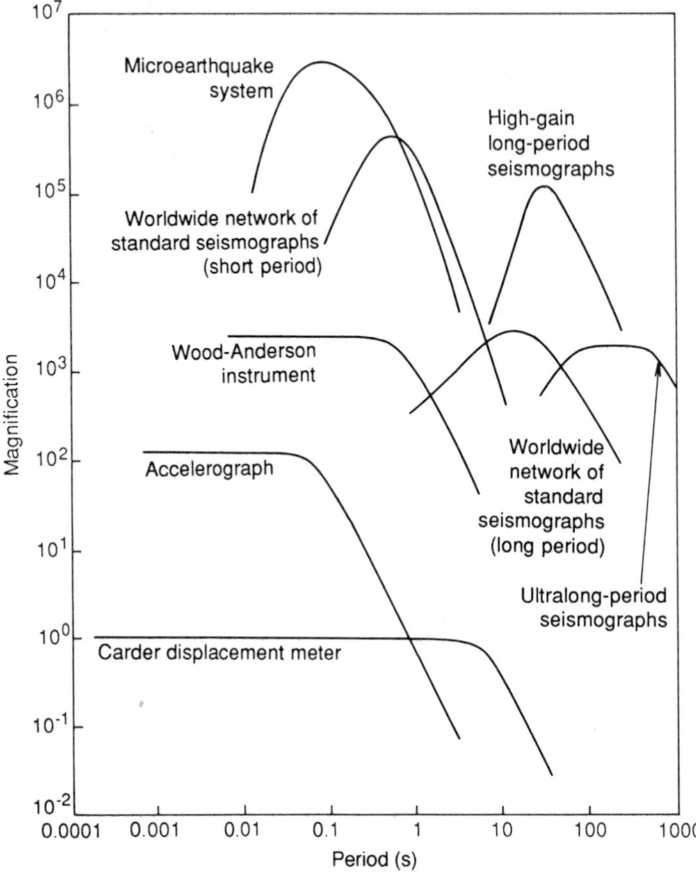

Figure 2.14 The magnifications of seismographs of different types and at different periods, i.e. the record amplitude divided by the ground amplitude. (After Boore, 1977)

experienced by an up-travelling wave drops steeply at the air–ocean interface and reflection is almost total ($R \sim -1$).

2.6 DIFFRACTION

So far we have assumed that interfaces between layers are flat and continuous. However, where changes occur in the interface over distances shorter than about a wavelength, or the dominant wavelength, diffraction effects occur. A common and important case is that of faults or sudden changes in the depth of a reflector (Figures 2.16–2.18) (Trorey, 1970). The fault edges act as point or linear sources producing hyperbolic type arcs in the plot of the reflection data (Figure 2.19) which look like anticlines, confusing the seismic section. One of the aims of seismic migration in data processing of reflection data is to remove such diffraction arcs by focussing them back to a point (Chapter 20).

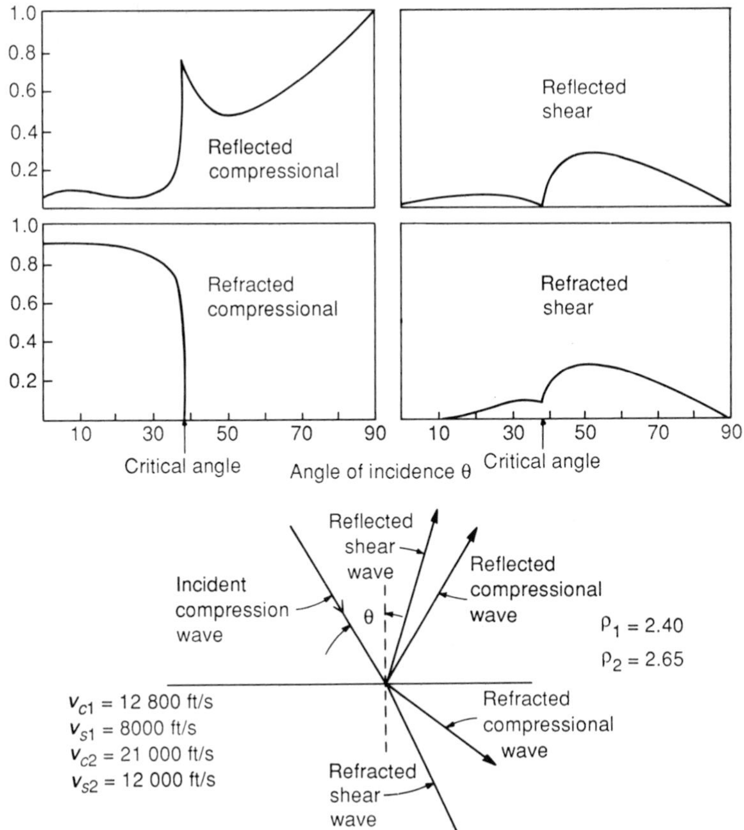

Figure 2.15 Partition of energy of incident P waves at a boundary for the velocities and densities illustrated. When a ray is not incident perpendicularly to the boundary, four rays result, reflected P and S, and refracted P and S. The amplitudes depend on the angle of incidence. (After Richards, 1961)

2.7 ABSORPTION AND ATTENUATION

As a body wave (P or S) spreads through the interior of the Earth the energy per unit area of the wave-front decreases inversely as the square of the distance. Since the energy of a wave is proportional to the square of the amplitude (in displacement or pressure), the amplitude of a body wave falls off inversely as the distance increases ($1/r$), plus some extra reduction where velocity increases with depth. However, for a surface wave, which is travelling in two dimensions over the Earth's surface, not three, its amplitude decreases as the inverse of the square root of the distance ($1/r^{1/2}$).

As well as this geometrical attenuation in amplitudes there is the physical absorption of seismic waves in the Earth (Figure 2.20), i.e. the conversion to heat energy in the series of compressions or shears (Frankel, 1991). This is quite small, particularly for low frequencies as in surface waves and the loss of energy per cycle is fairly constant for a large range of periods (100–1000 s). Thus the absorption increases with frequency and the higher frequencies are lost with increasing

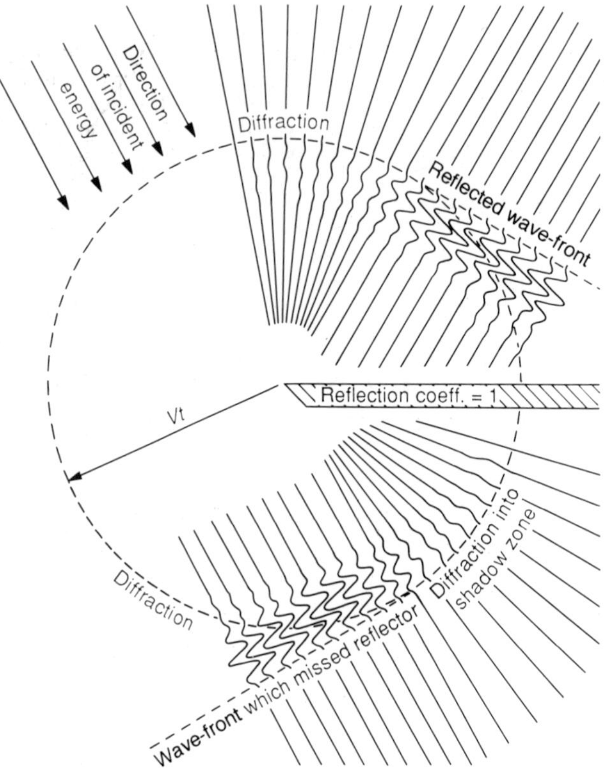

Figure 2.16 Diffraction, showing the effect of a plane wave incident on the edge of a barrier and the resultant amplitudes. (After Chevron Oil Co)

distance, or depth in the case of reflection seismology. There is also the effect of dispersion and some scattering from inhomogeneities. Thus the Earth acts as a low pass filter and the low-frequency surface waves dominate at most distances (Toksoz and Johnston, 1981; Cormier, 1982).

The absorption is caused by imperfect elasticity. This 'anelastic' damping is measured by the absorption coefficient a in the factor exp($-ar$) where r is the distance, or by the Q factor. Q is comparatively constant for different frequencies.

$$\frac{1}{Q} = \frac{\mathrm{d}E}{2\pi E} = \frac{av}{\pi f}$$

where dE is the energy loss per cycle, v the velocity and f the frequency.

The value of a is about 0.1–0.4 per km for granites and higher and more variable for sedimentary rocks (see Clark, 1966; Christensen and Wefner, 1989). Q can be regarded as the non-elastic response of the Earth for 'high' frequencies and as viscosity at 'low' frequencies. The data cannot distinguish between the different theories produced to explain seismic absorption, often described as internal friction, however, either dislocation mechanisms or grain boundary mechanisms (crystal defects), or both, are likely to be responsible (Karato and Spetzler, 1990).

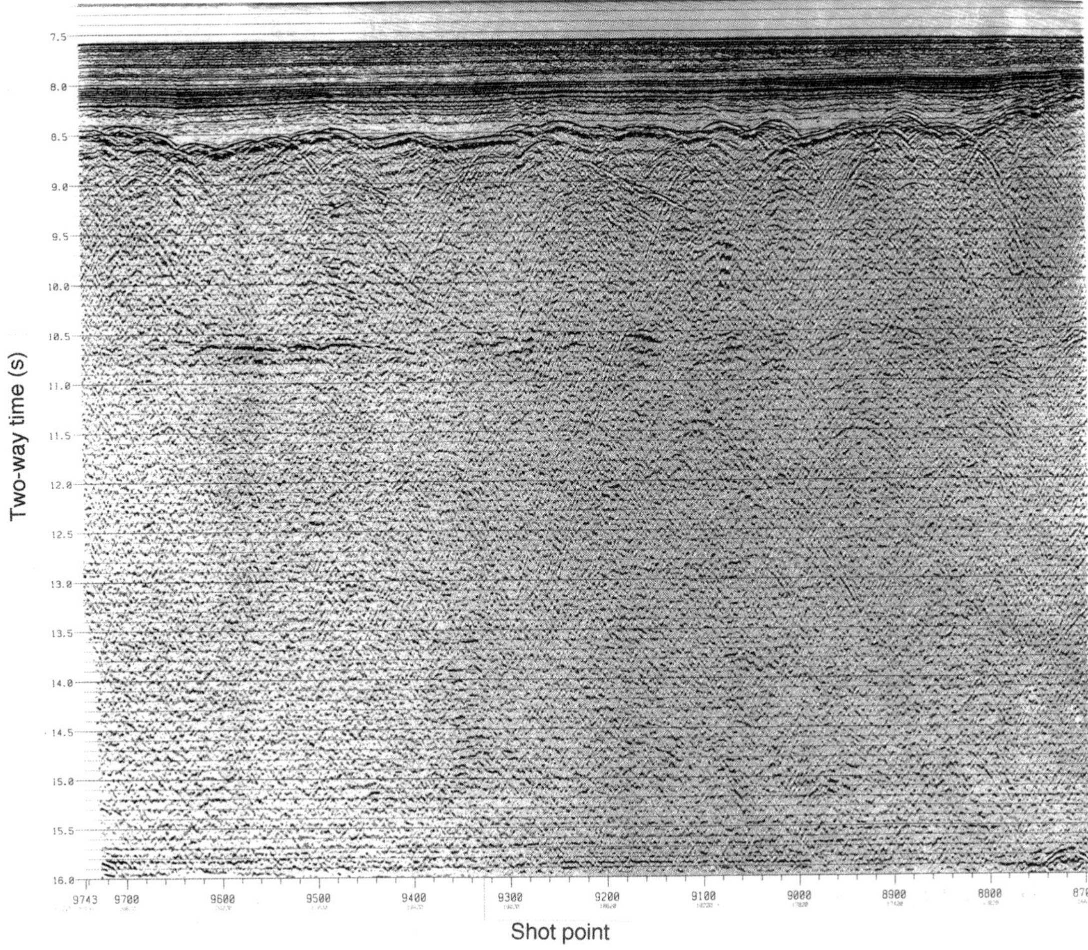

Figure 2.17 A reflection profile showing flat-lying sediments over basement (igneous) rocks whose rough surface produces diffraction effects. (Courtesy of Sharon Anker and Aust. Geol. Surv. Org.)

The attenuation of amplitudes with distance due to absorption, etc., varies with geological structure, being less in shield and platform regions, such as eastern North America, than in younger terrains, such as the western USA (Der *et al.*, 1982). Attenuation increases with the rock temperature and is greatest in the Earth's liquid outer core and in the low-velocity region of the upper mantle, more or less coincident with the low-velocity zone, suggesting high temperatures and possibly partial melting, as also indicated by S time delays (Hales and Doyle, 1967). Small ocean basins behind island arcs also have high attenuation (low Q), while cold subducting plates do not. Attenuation would be stronger near an earthquake focus. On the Moon attenuation is very low as it is a cold body.

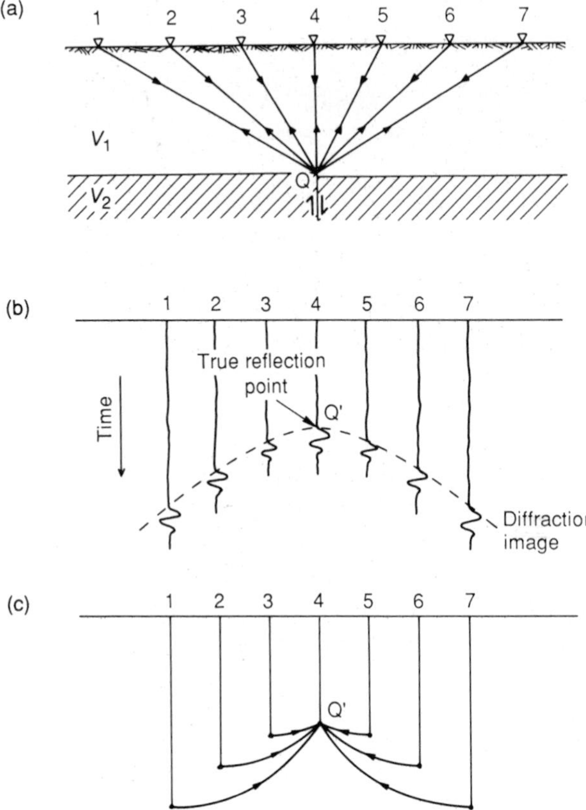

Figure 2.18 (a) Ray paths for diffraction from Q on a reflector at a vertical step. (b) The diffraction plotted without migration. (c) The diffraction plotted with migration to the true diffractor position. (After Robinson and Coruh, 1988)

2.8 MICROSEISMS AND NOISE

Seismic recording is affected a great deal by the amount of local seismic 'noise'. Seismograms show (Figure 1.3) that the ground is in continual motion with amplitudes of the order of micrometres (10^{-6} m) (Brune and Oliver, 1959; Fix, 1972; Murphy and Savino, 1975). Ground motion of frequency 1–10 Hz is produced by winds, local traffic, trains, etc., and slower vibrations by atmospheric low-pressure (cyclonic) systems over the oceans which produce *microseisms*, particularly in the winter months (Rind and Donn, 1978).

This microseismic motion of a few seconds period (maxima 6–8 s) is produced by the transfer of energy from the atmosphere to the ocean and thence to the ocean floor. Resonant standing waves on the sea, produced by interference between waves from winds of different directions in the cyclonic system, cause second-order pressure fluctuations on the sea floor (Longuet-Higgins, 1950). This energy is then transmitted through the crust for very long distances (10^3 km) as microseisms in the form of surface waves mainly, and can be detected even in the centre of continents. There are two frequency peaks in the

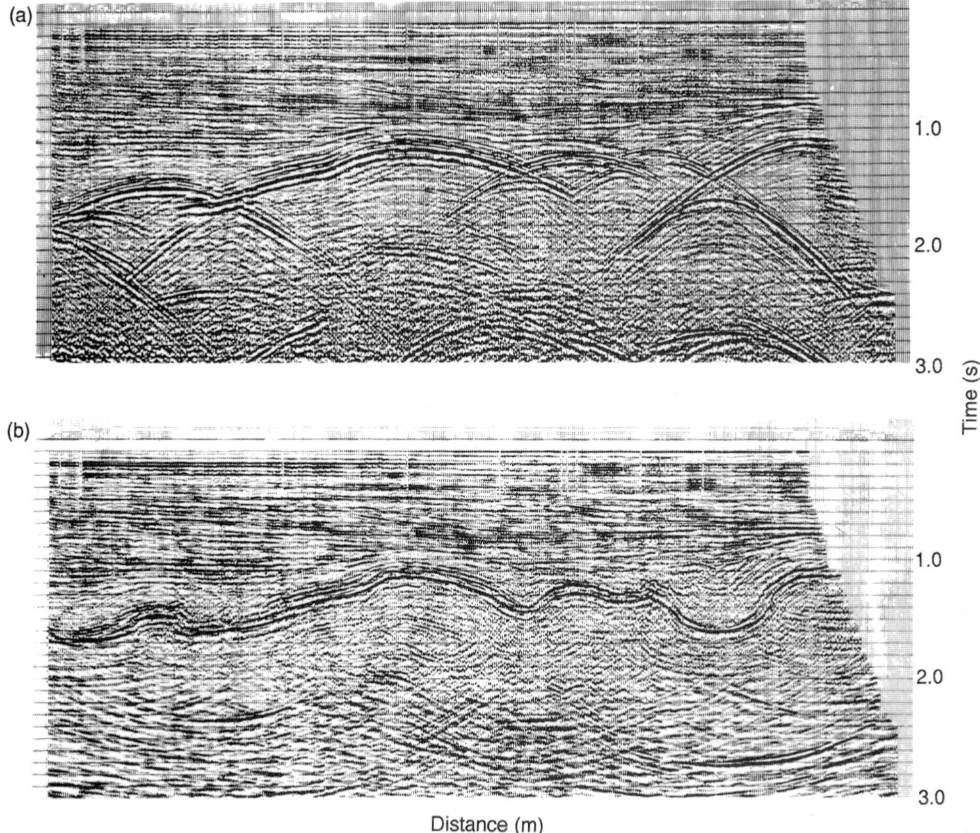

Figure 2.19 The effect of good migration: (a) unmigrated section showing the bow-tie effect; (b) the migrated section. (Courtesy Geco-Prakla)

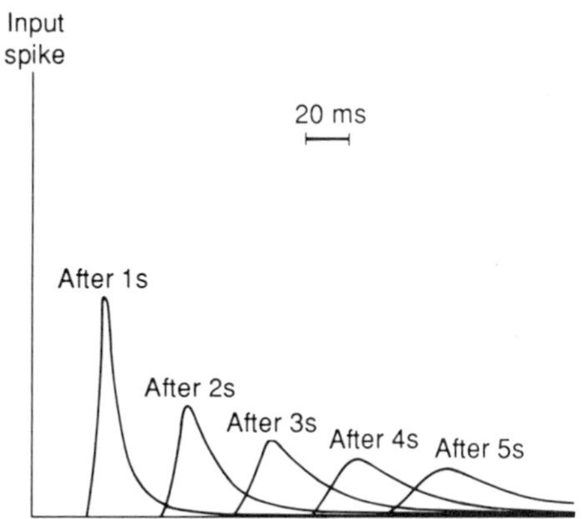

Figure 2.20 The effect of ground absorption on a seismic impulse, showing the change of shape and lengthening with travel time. (After Anstey, 1977)

microseisms, one at about 12 s, the period of ocean waves, and a larger peak at about 6 s (double frequency) due to the standing wave effect (Figure 2.21). Seismographs near the coast are also affected by waves breaking on the shore, but this does not explain the microseisms recorded far inland, which travel thousands of kilometres. One

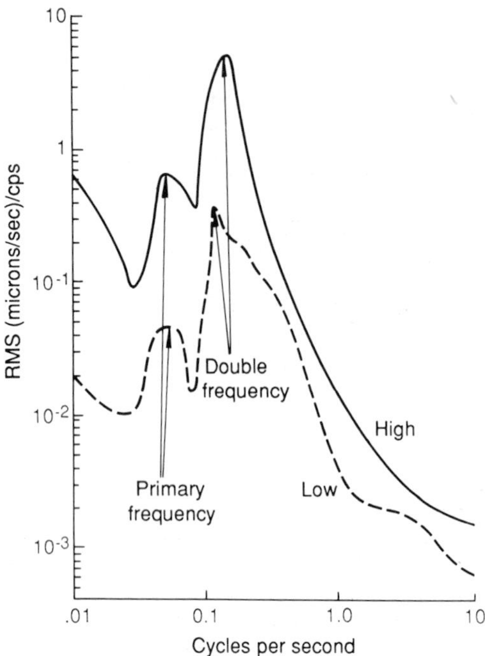

Figure 2.21 The spectra of amplitude spectral density versus period for vertical Earth motion showing the microseismic peaks. (After Iyer, 1970)

advantage of recording on the Moon is the lack of atmosphere and ocean, which results in very low seismic noise.

Noise is a perennial problem in seismic exploration and special techniques such as electrical filtering, deconvolution and redundant (multiple) recording are used to reduce it (see pp. 143 and 145). Noise reduces with depth being commonly surface waves, guided waves in near-surface layers and multiple reflections in shallow layers (Carter *et al.*, 1991).

Chapter 3
The seismograph

3.1 THE SEISMOGRAPH

The seismograph is as important to seismology and earth science as is the telescope to astronomy. It extends our 'hearing' far below audible frequencies and enables us to 'X-ray' the Earth. Seismic waves are recorded today over wavelengths from about a metre to thousands of metres, a ratio of about a million. However, for many years instruments were much narrower in frequency range.

The WWSN global network of stations, set up in the 1960s (Figure 1.6), is being gradually replaced by the new Global Digital Seismograph Network (GDSN) and the US National Seismograph Network, including a number of wide-band and digitally recorded instruments, some in boreholes 100 m deep to reduce noise (Peterson and Orsini, 1976; S. W. Smith, 1986). Broad-band seismograms of body waves from instruments that have flat responses for displacement and velocity in at least the frequency range 0.01 to 5.0 Hz have sufficient spectral information to enable estimates of the radiated energy of teleseisms, the associated stresses and the frequency properties of earthquake sources (Choy, 1990). The improved instruments have also made possible the development of waveform analysis and seismic tomographic imaging.

Very broad-band seismographs are based on a leaf-spring instead of a pendulum and have a flat response from 10 Hz to about 60 min, a remarkable range. The French Geoscope programme is to have 25 three-component digital broad-band instruments. US groups are installing ocean-bottom seismographs to help fill in gaps in the global coverage (Jacobson *et al.*, 1991). Some seismic stations now transmit data immediately to central offices via satellite (Romanowicz and Dziewonski, 1987). It is claimed that a special station BRV in Kazakhstan can detect earthquakes worldwide down to magnitude 4.2. Magnitude 4.5 or higher would be more usual.

Accelerometers (to study the effects of vibrations on buildings; see p. 75) and recording gravimeters have been installed in one global network (IDA), with 3000 accelerometers in the USA alone. A programme of studies (Passcal) using seismic arrays for Earth structure studies is also planned.

Today there are nearly 3000 seismographs around the globe, with about 1500 seismographs in the US alone, most reporting arrival times and preliminary interpretations to world seismic data centres such as the National Earthquake Information Centre of the US Geological Survey at Denver, Colorado (Masse and Needham, 1989), and the International Seismological Centre (ISC), Newbury, UK. About 10 000 earthquakes are located each year by the USGS, which publishes preliminary epicentral positions, depths, times and magnitudes weekly. More complete lists, including estimates of fault motion and earthquake moment, are published a few months later by the USGS, and final detailed listings in the bulletins of the ISC in due course. Interactive computer programs are used to sort out and separate the many reported 'phases' from different stations and to calculate the earthquake positions, depths and times. Quick epicentre information can be obtained within a few hours of a major event by dialling a computer at the National Earthquake Information Service of the USGS.

The ideal seismograph would record the ground motion relative to a stationary point above the ground. The best we can do in practice is to measure ground motion relative to a mass connected by a spring (Figure 1.4). When the ground moves the mass tends to remain stationary in two instances, initially, and when the vibration is very rapid relative to its natural frequency. This is the principle of the inertial seismograph. A coil and a magnet are attached to the moving mass and the fixed frame of the instrument respectively, or vice versa (Figure 1.4). The same principles apply in exploration seismology where the seismometer (the detecting element) is the geophone (Figure 3.1). In most instruments the relative motion of a coil and magnet provides a small voltage and current which can be amplified and digitally recorded or drive a pen recorder (or galvanometer and

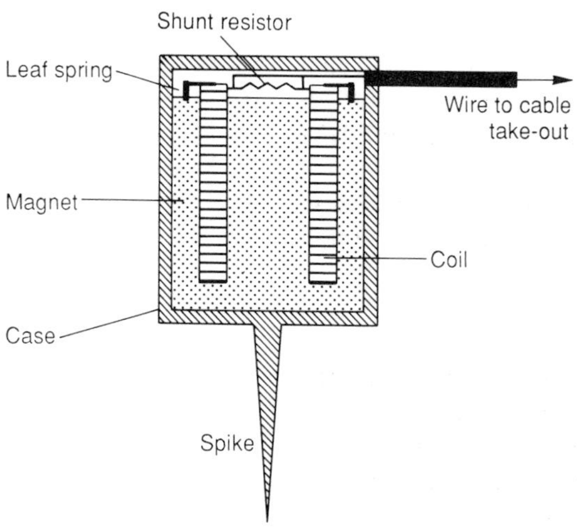

Figure 3.1 Schematic cross-section through a moving-coil geophone. (After Burger, 1992)

Figure 3.2 The seismic centre at the University of Alaska Geophysical Institute in Fairbanks, Alaska. The data are recorded digitally on a mainframe and on a PC computer and displayed on the recording drums. (Photo by James Kocia. Courtesy of the US Geological Survey)

light spot in older photographic recording instruments). Note the observatory recording drums in Figure 3.2.

The response of a seismometer is commonly proportional to the ground velocity because it depends on the relative velocity of the magnet and coil. The seismometer (and recorders such as pen systems and galvanometers) must be damped to give a reasonably flat response curve over the necessary frequency range (Figure 3.3). This stops instrument resonant vibrations which would swamp the records. Damping is usually obtained by an appropriate electrical resistance in parallel with the device which absorbs the unwanted vibrational energy (Figure 3.1).

The natural periods (the inverse of frequency) of seismometers and geophones and the frequency responses of the recording systems are designed for the ground wave frequencies of interest, e.g. in general seismology, for P, periods of 0.1 to a few seconds; for S, 0.5 to 15 s; and for surface waves, 15 to several hundred seconds (Table 3.1). In seismic exploration much smaller resolution is required because smaller targets and higher frequencies are involved, say 10–100 Hz or even 1000 Hz (and shorter periods and wavelengths), emitted by the small explosions and other energy sources used.

The natural period T of the seismometer or geophone depends on the inertial mass m of the moving magnet (or coil) and the spring constant k (its stiffness).

$$T = 2\pi(m/k)^{1/2}$$

The seismograph system also acts as a band-pass filter, the seismometer or geophone cutting off low frequencies because at lower frequencies the mass can follow the ground motion and there is less or no relative motion between the coil and magnet. On the other hand, recording systems have a high frequency cut-off.

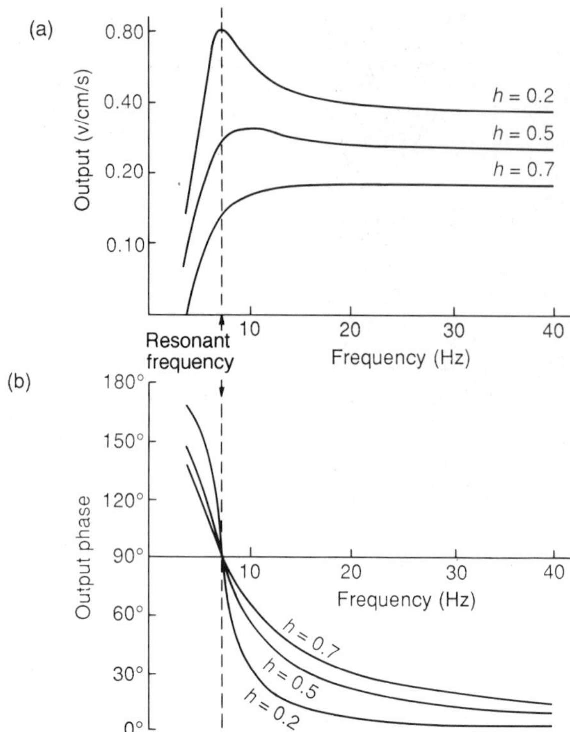

Figure 3.3 Responses of a geophone of natural frequency 7 Hz, for different damping factors *h*. Responses are for amplitude and phase relative to input phase. (After Telford *et al.*, 1990)

Important also is the timing on the seismograms provided by crystal clocks, as the most useful data in seismology are the arrival times, measured to at least 0.1 s on short-period earthquake records and to milliseconds in seismic exploration. A seismological observatory normally has three-component seismographs (i.e. a vertical and two horizontal, N–S and E–W) in both short and long periods; however, the new broad-band instruments may change this. The short-period instruments are used to record P, S and local earthquakes (which have higher frequencies) and the long-period instruments are used for distant earthquakes and surface waves. The peaks in the microseismic spectrum at 4–6 s are avoided by the short-period instruments peaking at about 1 s and the long-period 10–20 s (Figure 2.14). Broad-band seismographs with digital recording, large-amplitude (dynamic) range and electronic filtering will replace many dual short- and long-period instruments. The output from broad-band digital instruments is filtered to provide both short- and long-period data separately and to reduce noise. Broad-band seismographs provide frequency ranges greater than that of the ground motion for the first time, certainly for long distances.

The seismic spectrum extends from the slowest Earth vibration of almost an hour, to about 100 Hz near epicentres and explosions. At higher magnitudes more seismic energy is radiated at longer periods (and vice versa) because of the larger faults involved. Similarly, with

Table 3.1 The Seismic Spectrum (Some Common Values)

	Period	Frequency	Wavelength
Small explosion	0.01 s	100 Hz	30 m
Traffic, etc.	0.1	10	300 m
P (earthquakes)	1.0	1.0	5 km
S (earthquakes)	10	0.1	30 km
Surface waves	100	0.01	300 km
Earth oscillations	1000	0.001	3000 km
Earth tides	12 and 24 h	(see Figure 2.12)	
Earth rotation	24 h		

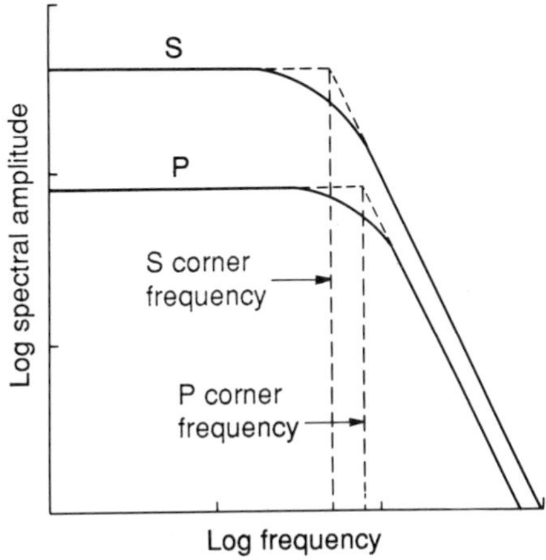

Figure 3.4 Idealized form of the body-wave spectra of an earthquake, corrected for propagation effects. (After Hanks and Wyss, 1972)

explosions the spectrum of the seismic pulse shifts to longer periods as the energy and volume of the source increase. Wavelengths vary from thousands of kilometres from distant earthquakes to tens of metres in seismic exploration (Table 3.1).

The spectra of earthquake waves fall off at higher frequencies, rolling over at the corner frequency (Figure 3.4). The corner frequency is useful for estimating the source dimension *r* (fault size), as Brune (1970) showed for a circular fault, one case which can be handled theoretically; *r* is also approximately half the length of a rectangular fault. The stress drop that occurs in an earthquake can also be estimated from the seismic moment M_0 (p. 47). Stress drops so calculated have been low compared to the total stresses thought to be present from direct measurements of crustal stress.

In both earthquake and seismic exploration work, digital recording on magnetic tape is replacing or has replaced the old analogue methods (e.g. see p. 143 and Dobrin and Savit, 1988, p. 68). This allows greater flexibility in data handling, such as filtering, and greater dynamic range.

An important development in general seismology in the 1960s was the use of arrays of seismometers recording at a central point (Figure

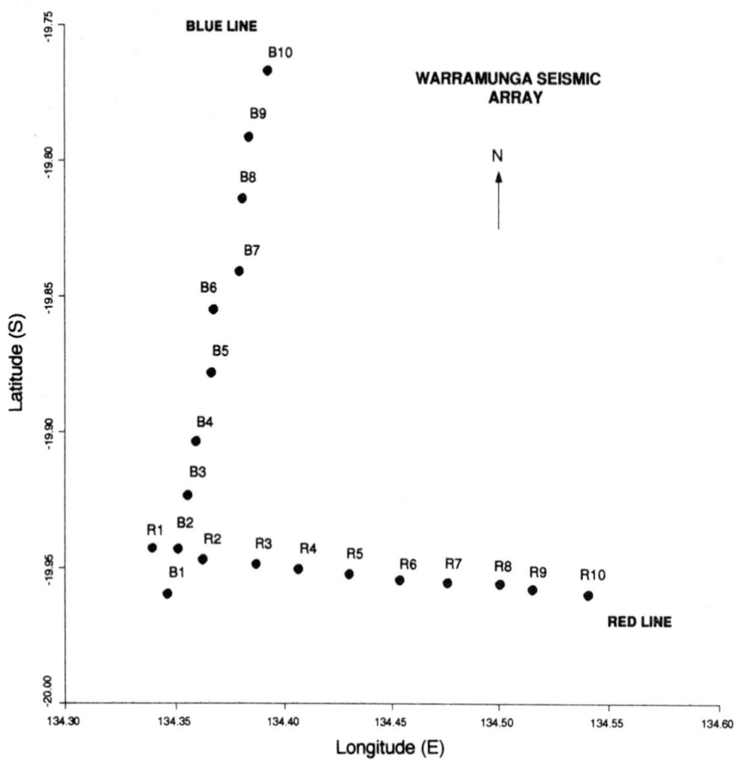

Figure 3.5 The seismograph array at Warramunga, Australia. Latitude and longitude are shown. (Courtesy Director, Aust. Geol. Surv. Org.)

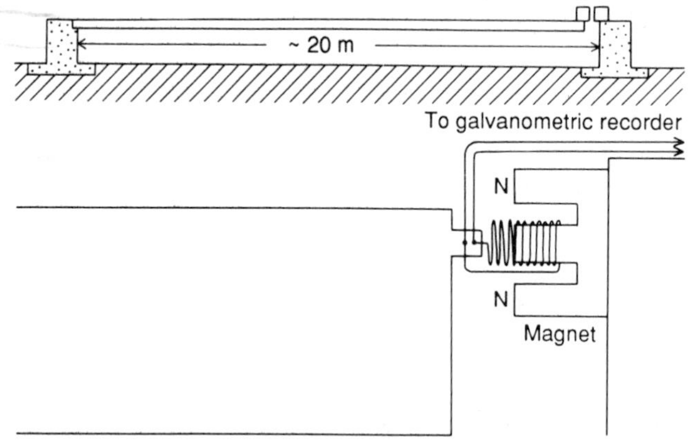

Figure 3.6 The Benioff strain seismograph; piers perhaps 20 m apart and electromagnetic sensors in the gap

3.5), as in exploration seismology but on a bigger scale. Such arrays allow accurate measurement of the difference in arrival time at the seismometers and so determination of the direction of arrival of phases (and of the ray parameter $p = (\sin i)/v$) by summing the amplitudes from the detectors with various time delays (see Gubbins, 1990).

An unusual seismograph is the strain seismograph designed by Benioff (1959; see also Agnew, 1986). It measures the strain (minute

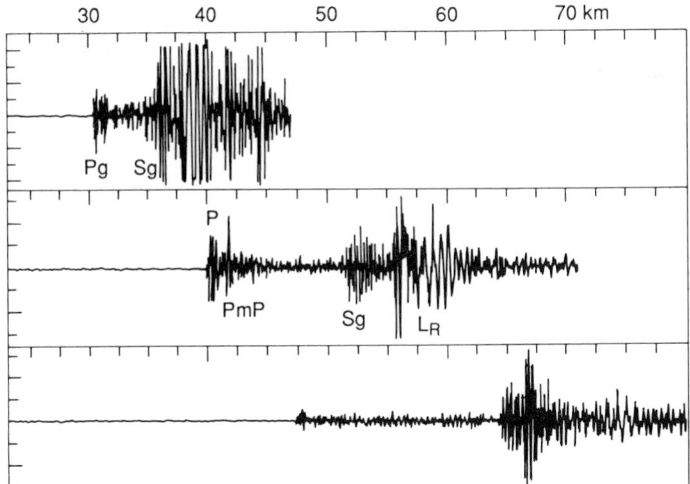

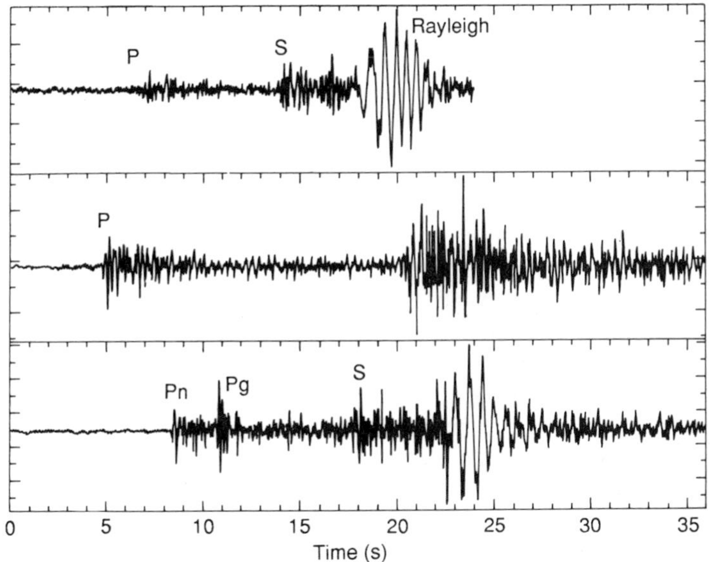

Figure 3.7 A local seismogram. Pg and Sg are upper crustal waves, PmP a reflection from the base of the crust, Pn a refraction from the top of the mantle, and L_R a Rayleigh wave. (Courtesy of Victor Dent and the Director of the Aust. Geol. Surv. Org.)

proportional changes in length) between two piers, say 20 m apart, by laser beams, or electromagnetically (Figure 3.6). It is useful for recording long-period oscillations of the Earth and the daily and semi-diurnal tidal deformations of the ground (Figure 2.12).

3.2 LOCATING EARTHQUAKES

Millions of earthquakes occur each year but most are too small and remote to be detected. The USGS alone locates 10 000 or more

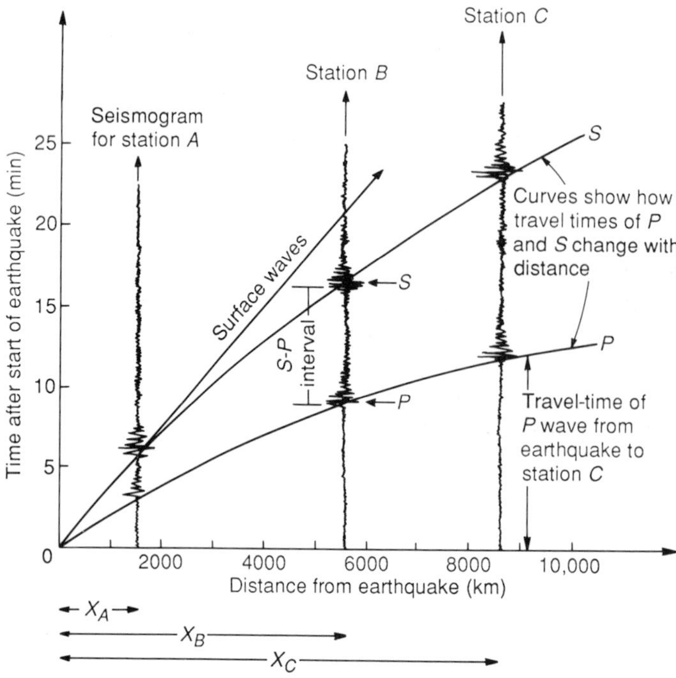

Figure 3.8 The distance of an earthquake (or explosion) from a seismograph station can be estimated by the difference between S and P arrival times. Tables of S–P times can be used for manual estimates (see Table 2.3). (After Press and Siever, 1986).

earthquakes each year with magnitudes (p. 45) ranging from 1 to 8 or more, using worldwide data. An earthquake may be initially located by comparing the *differences* in arrival time of various phases with standard travel-time tables and curves, such as the Jeffreys–Bullen tables or the new versions (see p. 18). Particularly useful at this stage are the S–P times for determining distance (see Figures 3.7 and 3.8; Tables 2.2 and 2.3; Simon, 1981; Kulhanek, 1990).

With three or four stations or more, arcs can be drawn on a map for near events, or on a globe for distant events. Depths may be estimated by the pP or sP times (see p. 44). For worldwide data, final epicentral locations and depths are determined at international centres by computers, using mainly the P arrival times at many stations worldwide and standard models of Earth structure, velocities and travel-times (Buland, 1976; Chen and Molnar, 1983). As the P arrivals are the first they are the most accurately read, not being in the noise of previous phases. The computer programs minimize the differences between observed and calculated times for different locations of the focus and for various origin times. Computers are also used to automatically pick the time of phase arrivals in some networks (Allen, 1982).

Even a single station recording can allow an approximate location if there is a clear S–P time and the ratio of the amplitudes of horizontal components can be used to give a direction. If the direction of the first P motion is clear there is no ambiguity in direction. As the waves are refracted upwards towards the surface, when the first P motion of the ground is upwards on the vertical component record, it indicates a push away from the epicentre towards the station, and vice versa.

In study of local seismic zones, such as in California, dense networks of seismographs allow focal determinations within the network as accurate as ± 0.5 km and depths to 1 km. However in worldwide studies mislocations of tens of kilometres can occur, particularly at subduction zones, because of structural and velocity inhomogeneities and uneven station distribution. The goal of one group is to determine the location and magnitude of large events in a local region within 10 s or so of its occurrence, enabling some warning to be given. One suggestion to obtain denser networks is for the use of privately owned and run seismographs which can send data by telephone lines to data centres (Cranswick *et al.*, 1993).

Chapter 4
Earthquake properties

4.1 DEPTHS

Earthquake depths are the least accurately determined value. They are estimated initially by the arrival time of reflected waves from the surface above the focus. These phases are denoted by pP and sP (Figure 1.2) and arrive seconds after the direct P wave at a time increasing with depth.

Most earthquakes are of 'shallow' depth (Figure 4.1), i.e. within the upper cooler crust (Sibson, 1982), and in the most brittle part of the lithosphere, particularly in continental regions and oceanic regions away from plate boundaries. This 'seismogenic zone' is 5–20 km deep in continental crust, with temperatures of 100–350 °C. The limiting higher temperature is usually explained as being due to the onset of plastic (ductile) creep. Intraplate (inside plates) oceanic earthquakes occur generally above the 750 °C isotherm, the higher temperature consistent with the transition from brittle to ductile behaviour for dry olivine rheologies (Chen and Molnar, 1983; Giardini, 1988).

In 1922 Turner suggested that some earthquakes occur considerably

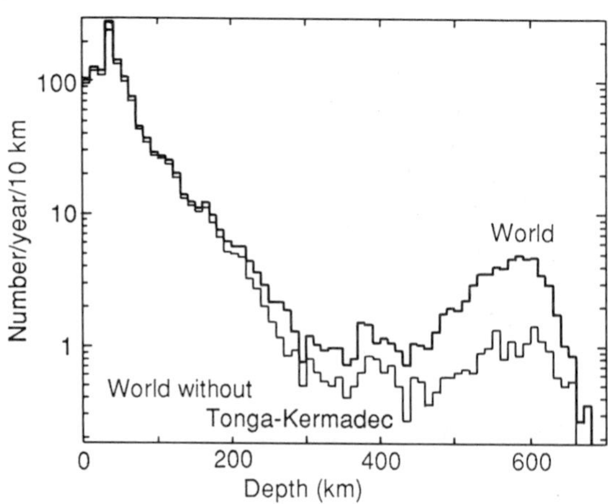

Figure 4.1 The number of earthquakes per year at 10 km depth intervals (1964–1986) with m_b of 5.0 or more. The lower line excludes the Tonga–Kermadec region. (After Frohlich, 1989)

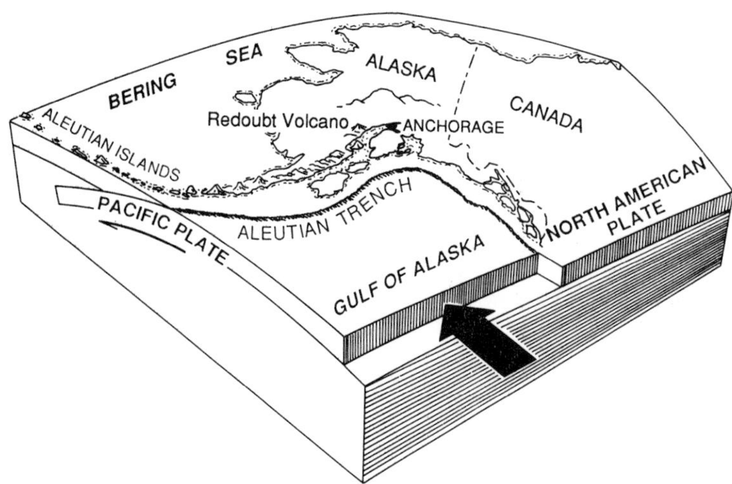

Figure 4.2 The Pacific plate thrusts under the North American plate along the Aleutian trench. (USGS, Circular 1061, with permission)

deeper and this was verified by Wadati in 1928. Much later it was shown that about 20% of earthquakes are deep events in the lower portion of subducted slabs (e.g. Figure 4.2), the Wadati–Benioff zones (Frohlich, 1989, 1994; Green, 1994) which extend under island arcs and continents, in some to depths of almost 700 km, with decreasing earthquake frequency. Figure 4.1 shows the decrease in earthquake frequency with depth and the minimum at about 300 km depth followed by an increase at 550–600 km and the cut-off near 700 km. Some 70% of the world's deep earthquakes occur in the Tonga Islands region, towards which the Pacific Plate is moving and subducting at a high rate. The mechanism of deep earthquakes has been investigated for many years as the great pressures at such depths are expected to stop ordinary fault movement (see p. 68).

4.2 EARTHQUAKE MAGNITUDE AND MOMENT

The magnitude of an earthquake is an instrumental measure of its size on a logarithmic scale introduced by Richter (1935; see also Boore, 1989), originally for near earthquakes recorded on a particular type of short-period seismograph, the Wood-Anderson. Modifications were later introduced so that the scale could be used on other recordings. A logarithmic scale is necessary because of the huge range of energy output by earthquakes (Table 4.1). The magnitude is normally estimated by measuring the ground amplitudes recorded at stations around the epicentre, perhaps worldwide, and allowing for distance, depth and wave period.

The general form of the empirical equation defining magnitude is

$$\mathrm{M} = \log_{10} A/T + F(D,h) + \text{constant}$$

Table 4.1 Details of earthquake magnitude

Magnitude (M_s)	Log M_0[a] (dyn. cm)	Log E[b] (ergs)	Approximate no. per year above M_s[c]	Approximate fault length (km)	Maximum intensity	Approximate maximum acceleration on hard ground ($g = 980\,cm/s^2$)
3	22.5	15.3	> 100 000		III	0.001g (minimum perceptible)
4	23.5	17.1	15 000		V	0.04g
5	24.5	19	3 000	4	VI–VII	0.1g
5.5	25.0	19.6	500	5–10	VII	0.15g
6	25.5	20.3	100	10–15	VII–VIII	0.2g
7	26.5	21.7	20	30–60	IX–X	0.6g
8	27.5	23.1	2	100–200	XI	1.0g

[a]Assuming $\log_{10} M_0 = M_s + 19.46$.
[b]Gutenberg and Richter (1956).
[c]Bolt (1993).

where A is the maximum ground amplitude in micrometres of the wave used, T is the wave period in s, F is an empirical function of D, the distance, in degrees, and h is the depth of focus in kilometres. The ground amplitude is calculated using the seismograph magnification at the appropriate wave period. A/T (or, Af, where f is frequency) is used because the energy density of the wave is required. For near earthquakes, magnitudes are also estimated from the durations of the recorded vibrations on short-period instruments (the length of the surface wave group).

Values on the original Wood-Anderson based scale are M_L magnitudes. But the international formula commonly used is for M_s the surface-wave magnitude for shallow earthquakes in the distance range 20–160°

$$M_s = \log_{10}(A/T) + 1.66 \log_{10} D + 3.3$$

in which the maximum vertical ground surface wave amplitude A is measured for periods T between 18 and 22 s. The M_s and M_L scales agree at about M_L 6 (Kanamori, 1983). For intermediate and deep events maximum short-period body-wave amplitudes, such as for P, PP and S are often used to give a *body* wave magnitude m_b, as introduced by Gutenberg. Different forms of the equation are necessary but the general form is $m_b = \log a/t + Q(D,h)$, where a is the maximum ground amplitude in micrometres, T is the period and $Q(D,h)$ is a depth and distance factor. Body waves recorded on digital broad-band instruments are also now being used to give spectral magnitudes.

Note that the magnitude scale, as defined, has no limits and can even be negative for very small events. In practice the scale saturates at magnitudes M_s above 8 or so when wavelengths exceed about 60 km, so that the magnitudes of very large earthquakes are underestimated (Kanamori, 1983). This is because such earthquakes have large rupture areas and radiate much long-period energy beyond that used in the magnitude measurements, the moment magnitude should then be used (see below). The largest known events have an equivalent magnitude of about 9.5. Magnitudes in some areas can be related to the length of fault rupture which varies from centimetres for a very small earthquake to about 1000 km for a very large one such as the 1960 Chile event. As a fault ruptures, seismic radiation occurs from a moving point or area and this can destructively interfere with the radiation from another part of the fault. The period at which this becomes detectable is used to estimate the dimensions of the fault.

4.2.1 Seismic Moment M_0

A different and more significant measure of earthquake size has come into use, the seismic moment $M_0 = \mu Ad$ (in dyne-cm), where μ is the

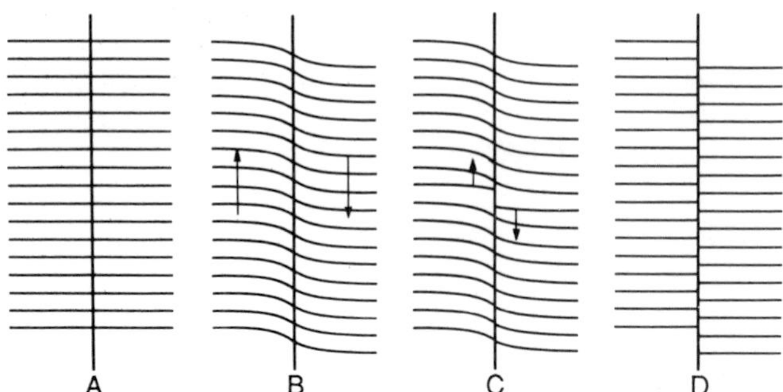

Figure 4.3 The 'elastic-rebound' model of earthquake mechanism showing the increasing strain and eventual fracture. (After Benioff, 1964)

rock rigidity (dynes/cm^2) or strength against shearing, A is the area of the fault movement, and d its displacement in cm. d and A can sometimes be estimated from field data and aftershock areas, but M_0 is usually estimated from the amplitudes of the long-period waves at large distances, corrected for attenuation, directional effects, etc. Thus M_0 is proportional to the level of the earthquake spectrum at low frequencies (Figure 3.4). If the area of the fault can be measured from aftershocks, but not the fault displacement, M_0 gives an estimate of the displacement.

The seismic moment refers to the two equal couples at a stressed fault, the causative stress shown in Figure 4.3 and the opposite elastic reactive couple (Aki and Richards, 1980). Seismic moment was introduced by Aki who showed that it is consistent with the earthquake slip and fault area determined from geodetic and geological data. Whereas magnitude is a purely empirical measure, the moment is independent of the type of fault. Hanks and Kanamori (1979) and Kanamori (1983) have, under certain assumptions, suggested a new magnitude scale, *moment magnitude* M_w, based on the seismic moment in which

$$M_w = \log_{10}(M_0/1.5) - 10.7 \qquad (M_0 \text{ in dyne-cm})$$

This scale has the important advantage that it does not saturate and the same formula can be used for shallow and deep earthquakes. For events smaller than 8, M_w agrees with M_s. The Chilean earthquake of 1960, perhaps the largest yet recorded, had an M_s value of 8.5, M_w 9.6, a seismic moment of 2.5×10^{30} dyn.cm and an energy release of 10^{26} ergs.

4.3 EARTHQUAKE INTENSITY

The intensity of an earthquake at a place is an *estimate* of the amount of shaking, building and other damage, human reaction, etc., and so varies from place to place but in general decreases with distance from the epicentre (Figure 4.4). Intensity scales are non-instrumental but are very useful indications of damage and its variation and distribution.

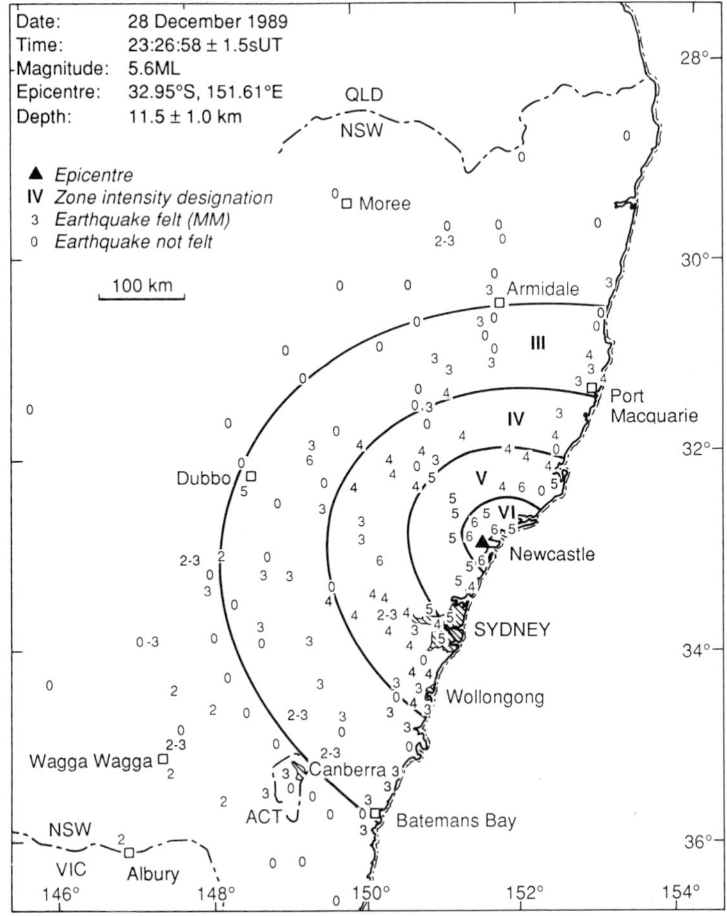

Figure 4.4 The isoseismals for the 1989 Newcastle earthquake, New South Wales, Australia. (Courtesy Director, Aust. Geol. Surv. Org.)

They have been in use much longer than the instrumentally measured magnitude scales. The scale most used to indicate effects on structures and people is the Modified Mercalli intensity scale; however, there are also Japanese and Russian scales. The Mercalli scale is defined from I to XII, using Roman numerals to distinguish it from magnitude scales (Table 4.2). Isoseismal contours of equal intensity are plotted for significant earthquakes as in Figure 4.4.

The intensities generated by a certain magnitude event depend to some extent on the geological region. The maximum intensity I_m near the epicentre and the magnitude of an event are related approximately for shallow shocks by the following relation (Richter, 1958) (other authors give different constants),

$$I_m = 3/2(M - 1)$$

This kind of relation is useful for estimating the likely effects of future earthquakes and the approximate magnitudes of historic ones

Table 4.2 Abridged modified Mercalli intensity scale

Intensity value	Description
I	Not felt. Marginal and long-period effects of large earthquakes.
II	Felt by persons at rest, on upper floors, or favourably placed.
III	Felt indoors. Hanging objects swing. Vibration like passing of light trucks. Duration estimated. May not be recognized as an earthquake.
IV	Hanging objects swing. Vibration like passing of heavy trucks; or sensation of a jolt like a heavy ball striking the walls. Standing cars rock. Windows, dishes, doors rattle. Glasses clink. Crockery clashes. In the upper range of IV, wooden walls and frame creak.
V	Felt outdoors; direction estimated. Sleepers awakened. Liquids disturbed, some spilled. Small unstable objects displaced or upset. Doors swing, close, open. Shutters, pictures move. Pendulum clocks stop, start, change rate.
VI	Felt by all. Many frightened and run outdoors. Persons walk unsteadily. Windows, dishes, glassware broken. Knickknacks, books, etc., off shelves. Pictures off walls. Furniture moved or overturned. Weak plaster and masonry D cracked. Small bells ring (church, school). Trees, bushes shaken visibly, or heard to rustle.
VII	Difficult to stand. Noticed by drivers. Hanging objects quiver. Furniture broken. Damage to masonry D, including cracks. Weak chimneys broken at roof line. Fall of plaster, loose bricks, stones, tiles, cornices, also unbraced parapets and architectural ornaments. Some cracks in masonry C. Waves on ponds, water turbid with mud. Small slides and caving in along sand or gravel banks. Large bells ring. Concrete irrigation ditches damaged.
VIII	Steering of cars affected. Damage masonry C; partial collapse. Some damage to masonry B; none to masonry A. Fall of stucco and some masonry walls. Twisting, fall of chimneys, factory stacks, monuments, towers, elevated tanks. Frame houses moved on foundations if not bolted down; loose panel walls thrown out. Decayed piling broken off. Branches broken from trees. Changes in flow or temperature of springs and wells. Cracks in wet ground and on steep slopes.
IX	General panic. Masonry D destroyed; masonry C heavily damaged, sometimes with complete collapse; masonry B seriously damaged. General damage to foundations. Frame structures, if not bolted, shifted off foundations. Frames racked. Serious damage to reservoirs. Underground pipes broken. Conspicuous cracks in ground. In alluviated areas, sand and mud ejected, earthquake fountains, sand craters.
X	Most masonry and frame structures destroyed with their foundations. Some well-built wooden structures and bridges destroyed. Serious damage to dams, dykes, embankments. Large landslides. Water thrown on banks of canals, rivers, lakes, etc. Sand and mud shifted horizontally on beaches and flat land. Rails bent slightly.
XI	Rails bent greatly. Underground pipelines completely out of service.
XII	Damage nearly total. Large rock masses displaced. Lines of sight and level distorted. Objects thrown in the air.

see Richter (1958) for definitions of masonry types

for which there is only intensity information. Intensity values also allow approximate estimation of ground accelerations, velocities and displacements, important in understanding effects on buildings (Trifunac and Brady, 1975, 1976). For example, they suggested the following relations for peak accelerations, velocities and amplitudes in California, but they can be useful elsewhere:

accelerations a for IV $\leq I_m \leq$ X	$\log a_v = -0.18 + 0.30 I_m$ (cm/s^2)
	$\log a_h = 0.014 + 0.30 I_m$ (cm/s^2)
velocities c for IV $\leq I_m \leq$ X	$\log c_v = -1.10 + 0.28 I_m$ (cm/s)
	$\log c_h = -0.63 + 0.25 I_m$ (cm/s)
amplitudes d for V $\leq I_m \leq$ X	$\log d_v = -1.13 + 0.24 I_m$ (cm)
	$\log d_h = -0.53 + 0.19 I_m$ (cm)

where v and h refer to vertical and horizontal components.

However, such estimates depend also on the local geology and soil, the type of faulting and attenuation in the region. Amplification factors for land on mud and artificial fill can be several times or more. Clear examples of the effect of local conditions occurred in the 1989 Loma Prieta, California, and the 1985 Mexican earthquakes (see p. 75).

Accelerations as high as 1–2g (g is the normal gravitational acceleration 9.8 m/s^2) have been recorded on accelerometers. Horizontal vibrations are the most damaging to structures, particularly at 0.05–0.1g and above. The lowest acceleration perceptible by people at a frequency of a few hertz is about $10^{-3}g$ or 1 cm/s^2. At sea, close earthquakes may be felt as an upward jolt of the ship because the P waves are refracted upwards by the low-velocity ocean.

4.4 EARTHQUAKE ENERGY

An increase in magnitude of one unit means a tenfold increase in amplitude; however, the seismic energy increases by 32 times. For example, a magnitude 8 earthquake is not twice the energy of a magnitude 4 event, but over a million times greater. This is because the energy is a logarithmic function of magnitude, $\log E = 11.8 + 1.5\,M_s$ ergs (Gutenberg, 1956). However, as we saw, the magnitude scale saturates for great earthquakes when the rupture dimensions exceed about 60 km, the wavelengths of the waves used in the magnitude determination (Kanamori, 1983), leading to underestimation of the energy. In terms of the moment scale earthquake energy is $M_0/(2 \times 10^4)$ ergs.

Thus earthquakes have a huge energy range, 15 orders of magnitude, from the size of a small rock-burst in a mine to an earth shaker of 10^{26} ergs (the Chile event of 1960), equivalent to a 300 or more megatonne nuclear explosion (1 ton of chemical explosive = 4.18×10^{16} ergs) (see Table 4.1 and Figure 4.5).

The average global seismic energy released per year is 10^{25} ergs (10^{18} joules), equivalent to 3×10^7 kilowatts, or one extremely large

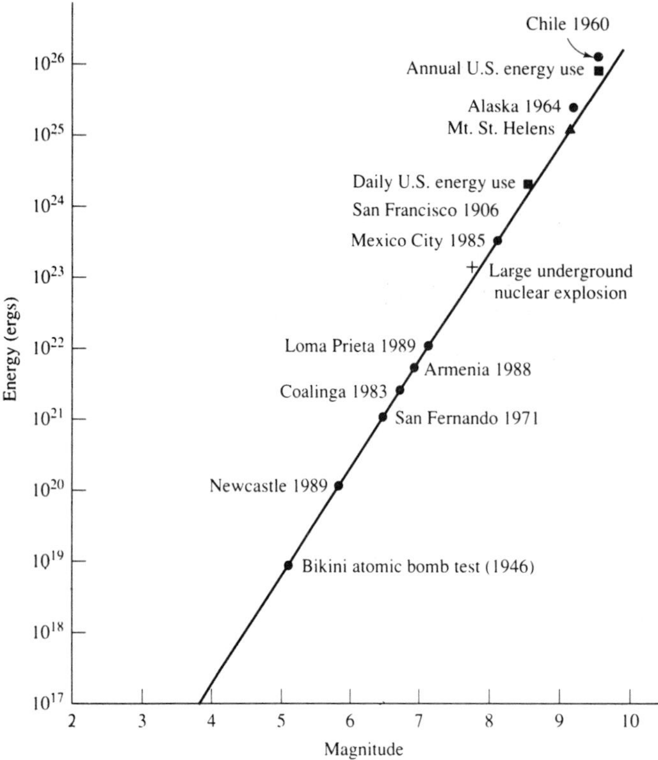

Figure 4.5 Energy versus the equivalent magnitude (M_W) of some natural man-made and seismic phenomena. (From Bolt, 1993, with permission)

earthquake per year. This is about one-thousandth of the global rate of heat flow from the interior, so that the seismic energy can be regarded as a by-product of the Earth's heat flow through the accumulation of stress in the lithosphere, related to mantle convection and plate tectonics. Note that most of the seismic energy release comes from the largest shocks, e.g. about 80% come from those at magnitude 7.9 and above (Table 4.1). Although there are many more small shocks than large ones, the small shocks are not very significant in releasing stress energy.

Chapter 5
Earthquake frequency and distribution (seismicity)

5.1 EARTHQUAKE FREQUENCY

Seismicity is the study of the distribution, frequency and magnitude of earthquakes around the globe (Young *et al.*, 1988). About 50 000 earthquakes are felt each year and many more are recorded instrumentally. In their pioneering studies of seismicity, magnitude and energy, Gutenberg and Richter (1956) showed that the frequency of earthquakes increases strongly as magnitude *de*creases, e.g. for worldwide shallow events and for many regions the following holds: $\log N = 8.2 - M$, where N is the number of earthquakes per year of magnitude above M. Thus there are about nine times more in a certain magnitude range than in the one above (Table 4.1). This is more generally expressed as $\log N_{10}(M) = A - bM$ in which the b value is close to 1 and A is the level of activity. However, data do not fit the relation well at either end of the magnitude scale and Pacheco *et al.* (1992) point out its inaccuracy (see also Frohlich and Davis, 1993). This magnitude–frequency relation, where true, may be because earthquakes are a fractal and stochastic process, occurring on a fractal network of faults. Seismogenic areas seem to be in a critical state triggered by random changes (Bak and Tang, 1989).

5.2 GEOGRAPHIC DISTRIBUTION

Earthquakes remind us of the active nature of the Earth and how dynamic it really is. They are involved in mountain building (Molnar and Chen, 1982) and outline the mobile plates (Figure 1.9) which are moving at the rate of centimetres per year, ultimately driven by convecting motions deep in the Earth (Cox and Hart, 1986; Gubbins, 1990; Keary and Vine, 1990; Wilson, 1993). Pacheco and Sykes (1992) have published a new catalogue of large ($M_s > 7$) shallow earthquakes

Table 5.1 Some important earthquakes

Date	Place	Magnitude (M_s)[a]	Moment magnitude (M_w)[b]	Deaths[c]	Effects (displacements usually maximum)
23 Jan 1556	Shensi, China			(830 000)	Many buried in loess, starvation
1 Nov 1755	Lisbon area	8 +		70 000	Great tsunami, seiches
4 Feb 1783	Calabria, Italy			50 000	
16 Dec 1811	New Madrid, USA	(8)		several	
23 Jan; 7 Feb 1812	New Madrid, USA	(8.3)			
23 Jan 1855	Wellington area, New Zealand	(8.1)			
9 Jan 1857	Fort Tejon, CA, USA				San Andreas Fault, 4 m rupture
Aug 1868	Arica, Peru (now Chile)				Large tsunami across Pacific
16 Aug 1868	Ecuador and Colombia			70 000	
26 Mar 1872	Owens Valley, CA, USA			(50)	7 m vertical; 5 m horizontal slip
31 Aug 1886	Charleston, SC, USA			(60)	
28 Oct 1891	Mino-owari, Japan			7000	4 m vertical; 7 m horizontal slip
15 June 1896	Riku-ogu, Japan			22 000	
12 June 1897	Assam, India	(8.7)		1500	11 m displacement
3, 10 Sept 1899	Yakutat Bay, AK, USA	(7.8, 8.6)			14 m vertical uplift
4 Apr 1905	Kangra, India	7.4	7.7	19 000	
9, 23 July 1905	Mongolia	7.5, 7.6	8.4, 8.4		
31 Jan 1906	Ecuador	8.1	8.5	1000	6.4 m horizontal; 1 m vertical; 330 km length
18 Apr 1906	San Francisco, CA, USA	7.7	7.9	700	
17 Aug 1906	Chile	7.7	7.7	20 000	
28 Dec 1908	Messina, Italy	7.5		58 000	
3 Jan 1911	Turkestan	7.7	7.7	450	Large displacements
13 Jan 1915		7		30 000	
1915	Avezzano, Italy	7.5		32 610	
3 Oct 1915	Pleasant Valley, NV, USA	7.5	7.5		
16 Dec 1920	Kansu, China	8.4	8.3	> 100 000	Major fractures
11 Nov 1922	Central Chile	8.1	8.7	600	
1 Sept 1923	Near Tokyo	8.0	7.9	99 300	Great fire, 4.5 m displacement
7 Mar 1927	Tango, Japan	7.4	7.1	3000	
22 May 1927	Tsing Hai, China	7.7	7.7	40 912	Large fractures
16 June 1929	Buller, New Zealand	7.4	7.4	17	5 m uplift
18 Nov 1929	Grand Banks, Atlantic Ocean	7.0	7.0		Undersea avalanches, cable breaks
2 Feb 1931	Hawke's Bay, New Zealand	7.6	7.6	256	1–2 m displacement
2 Mar 1933	Japan	8.3	8.4	3000	
10 Mar 1933	Long Beach, CA, USA	6.3		(100)	
15 Jan 1934	Bihar-Nepal	8.1	8.0	10 700	Large fissures

31 May 1935	Quetta, Pakistan	7.4	8.1	25 000	
25 Jan 1939	Chile	7.6	7.6	28 000	
26 Dec 1939	Erzincan, Turkey	7.6	7.6	32 700	3.7 m strike-slip
1948	Ashkabad, former USSR	7.3		19 800	
28 June 1948	Fukui, Japan	7.1	6.9	5130	
15 Aug 1950	Assam, Tibet	8.6	8.6	1526	Large topographic changes
4 Mar 1952	Tokachi-Oki, Japan	8.3	8.1	600	Tsunami
21 July 1952	Kern County, CA, USA	7.8	7.2	11	Considerable damage, >1 m vertical
4 Nov 1952	Kamchatka	8.2	9.0		
16 Dec 1954	Nevada, USA	7.1 and 6.8			4 m vertical, 4 m horizontal, 104 km long
12 Aug 1953	Ionian Isles	7.1	7.1	500	A long series of activity
9 July 1956	Santorini Is., Greece	7.7	7.7	57	Tsunami, volcanism
9 Mar 1957	Aleutian Is.	8.1	8.9		
28 July 1957	Mexico	7.5	7.7	55	
4 Dec 1957	Altai-Gobi, Mongolia	8.0	8.1	30	250 km rift
10 July 1958	Southern Alaska, USA and British Columbia, Canada	7.9	7.7	5	7 m strike-slip, Fairweather fault
18 Aug 1959	Hebgen L., MT, USA	7.5	7.2	28	
29 Feb 1960	Agadir, Morocco	5.9		14 000	Shallow, under the city
22 May 1960	Chile	7.8	6.9	5700	Largest yet recorded, 1600 km fault, tsunamis, landslides, volcanism
		8.5	9.5		
		8.5	9.5		
1 Sept 1962	NW Iran	7.3		12 230	
26 July 1963	Skopje, Yugoslavia	6.0		1200	80% destruction, shallow, near city
28 Mar 1964	Alaska, USA	8.4	9.2	178	6 m displacement, liquefaction causes landslides in Anchorage, tsunami
26 Apr 1966	Tashkent, former USSR	5.3			Under city, much damage, many aftershocks
16 May 1968	SE Hokkaido	7.7	7.7	48	
24 May 1968	Inangahua, New Zealand	7.1	7.1		Large landslides
31 Aug 1968	Iran	7.1	7.2	11 600	Surface faulting
14 Nov 1968	Meckering, Western Australia	6.9			37 km fault break
28 Feb 1969	Atlantic Ocean	7.8	7.8	2	Felt in Portugal, Spain, Morocco
31 May 1970	N Peru	7.6	7.9	66 794	Two towns buried in mountain slides
10 Jan 1971	W New Guinea	7.9	7.7		
9 Feb 1971	San Fernando, CA, USA	6.8	6.7	64	US $550 million damage
23 Dec 1972	Managua, Nicaragua	6.2		5000	Accelerations $>0.3g$
30 Jan 1973	Michoacan, Mexico	7.3	7.6		
4 Feb 1975	Liaoning, China	7.2	6.9		A successful prediction from many foreshocks
14 Jan 1976	Kermadec Is., Pacific Ocean	7.9	7.9		

continues overleaf

Table 5.1 (*continued*)

Date	Place	Magnitude (M_s)[a]	Moment magnitude (M_w)[b]	Deaths[c]	Effects (displacements usually maximum)
4 Feb 1976	Guatemala	7.5	7.6	22 000	200 km rupture
6 May 1976	Friuli, Italy	6.5		965	Extensive damage
27 July 1976	Tangshan, China	7.8	7.4	240 000	Great damage
19 Aug 1977	Sumbawa, Indonesia	8.1	8.3		Large tsunami
29 Nov 1978	Oaxaca, Mexico	7.6	7.6		Predicted
1978	Tabas, Iran	7.7		18 220	
23 Nov 1980	Southern Italy	7.2		> 3000	
2 May 1983	Coalinga, CA, USA	6.6			
19 Sept 1985	Michoacan, Mexico	8.1	8.0	9500	US $3 billion damage, mostly Mexico City
20 Oct 1986	Kermadec Is., Pacific Ocean	8.1	7.8		
1 Oct 1987	Whittier, CA, USA	5.9	6		$358 million damage in Los Angeles area
7 Dec 1988	Armenia	7.0		25 000	
23 May 1989	Macquarie Ridge	8.2	8.0		Long strike-slip movement
17 Oct 1989	Loma Prieta, CA, USA	7.1	6.9	62	US $6 billion damage in San Francisco and Santa Cruz, 1.9 m horizontal and 1.3 m vertical movement
27 Dec 1989	Newcastle, Australia	5.5		12	First certain known fatalities in Australia
1990	Iran	7.3		40 000	
16 July 1990	Luzon, Philippines	7.8		2000	Large fissures and landslides
28 June 1992	Landers, CA, USA	7.5		1	70 km surface rupture
17 Jan 1994	Northridge, CA, USA	6.7	6.7	33	Reverse fault under urban area, US $13–20 billion damage
16 Jan 1995	Kobe, Japan	7.2	6.9	> 5000	Strike-slip; 103 000 buildings destroyed
27 May 1995	Sakhalin, Russia	7.1		2000	Many houses destroyed

[a] M_s values largely from Pacheco and Sykes (1992) and Kanamori (1977).
[b] M_w values calculated from Pacheco and Sykes (1992) M_0 values, or from Kanamori (1977).
[c] From US National Oceanic and Atmospheric Administration; Tazieff, *When the Earth Trembles*, Harcourt Brace & World Inc. and Rupert Hart-Davis Ltd; Bath (1967).

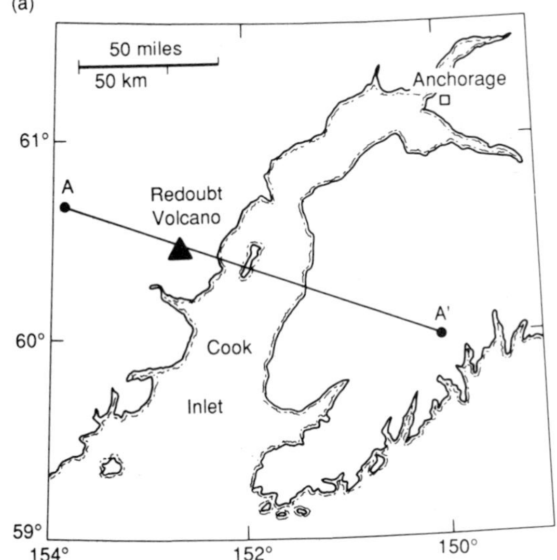

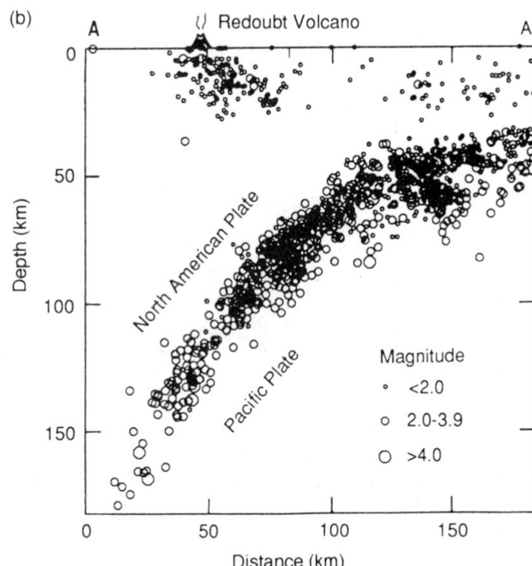

Figure 5.1 Earthquake foci under Cook inlet, Alaska, January 1980–November 1989. (a) The location of the cross-section A–A′ which is above the boundary between the Pacific and North American plates; (b) vertical section showing foci within 50 km of the section. A zone of shallow earthquakes lies beneath the active volcanoes west of Cook Inlet. (Courtesy USGS, with permission)

from 1900 to 1989 with calculated seismic moments and some of the more important earthquakes are listed in Table 5.1.

Ninety-five per cent of all earthquake energy is released at plate boundaries and only 5% inside the plates (intraplate). Some 86% occur at subducting plate margins (down-going Wadati–Benioff zones) where oceanic plates older than about 20 million years have cooled and become dense enough to sink into the mantle, such as at island arcs and along the west coast of South America and Central America and along the coasts of Alaska (Figures 4.2 and 5.1) and Japan. Subduction zones such as these have been associated with the largest earthquakes that occur, e.g. the magnitude 8 + earthquakes of Tokyo (1923), Chile (1960), Alaska (1964) and Mexico (1985) (see Table 5.1). The great Chilean earthquake of May 1960 had a moment magnitude of 9.5, the largest recorded this century. This event produced large changes in height over an area 1000 km long by 200 km wide, with a maximum uplift of 5.7 m at Guamblin Island and subsidence of 2.7 m in Valdivia. Plafker (1972) modelled this as a rupture 1000 km long, 120 km wide and dipping 20°E with 20 m of dip movement.

Some subduction zones show evidence of a double seismic structure (Figure 5.2), two planes of epicentres, the upper one being along the upper surface of the subducting slab. As focal mechanism studies (p. 70) show, the upper plane is in compression (Minoura and Hasegawa, 1992). The lower plane is in the middle of the plate and is apparently in tension as it drops down. There is some evidence for penetration of the mantle by the subducting plates to as deep as 1000 km (Creager and Jordan, 1986), well below the maximum depth of seismicity, or even to near the core–mantle boundary (Grand, 1994).

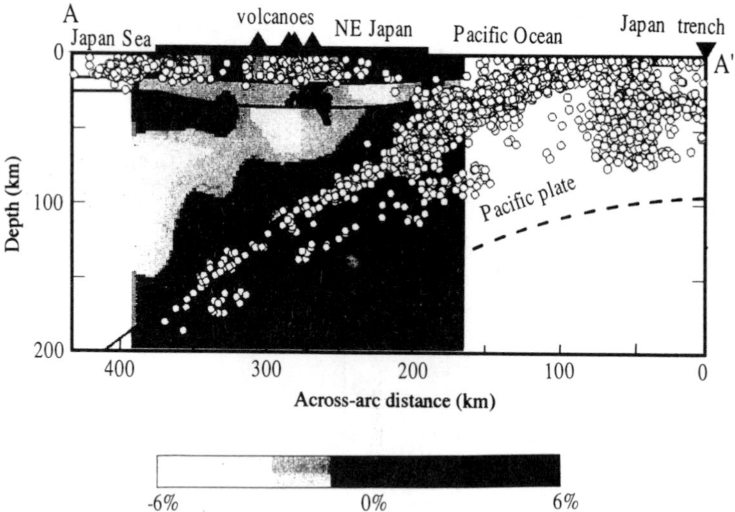

Figure 5.2 The double seismic structure where the Pacific plate subducts under Honshu, Japan. The shading shows velocity variations. (After Hasegawa *et al.*, 1994)

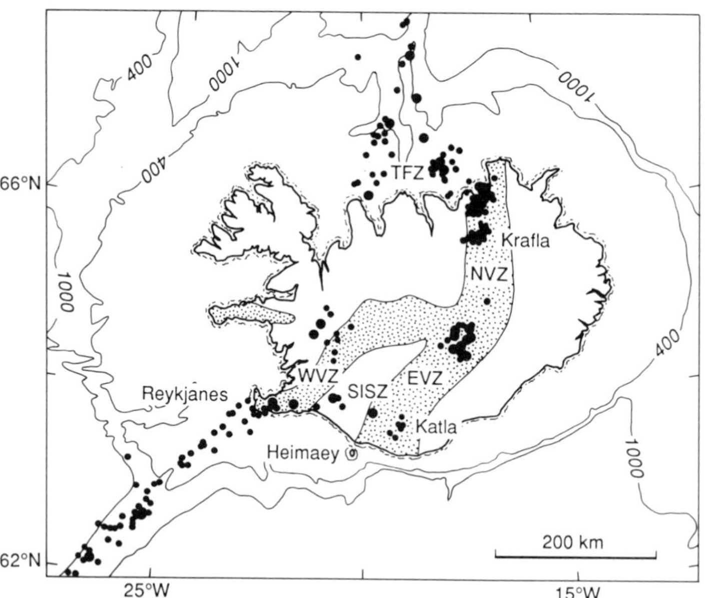

Figure 5.3 The Iceland eastern, western and northern volcanic zones (EVZ, WVZ, NVZ), the South Iceland seismic zone (SISZ), the Tjörnes fracture zone (TFZ) and the adjoining Atlantic ridges north and south. (After Einarsson, 1991)

Only 9% of the Earth's seismicity occurs at the mid-ocean ridges where new plate is formed by rising magma (at a rate of about 3 km^2/year). Recent velocity studies indicate that it is hot beneath the mid-ocean ridges to depths of 200 km or so, and colder generally under the continents. Iceland is an interesting case of an island built up by volcanism over a mantle 'hot spot' (Figure 5.3) and straddling the Mid-Atlantic Ridge (Einarsson, 1991). Mid-ocean ridges and hot spots are the important forms of convective upwelling from the Earth's interior.

Crustal deformation and plate motions are now being measured with great accuracy by using the radio signals broadcast by the US Global Positioning System (GPS) of satellites (21 plus three spares), and very long baseline interferometry (VLBI) using quasars as sources (Hager *et al.*, 1991; Gordon and Stein, 1992), plus satellite laser ranging. Under good conditions the distance between two stations thousands of kilometres apart can be measured to a few centimetres or even millimetres (Sato, 1993). Such space geodetic data show that plate velocities averaged over several years are similar to those from ocean magnetic data averaged over millions of years. Displacements in particular earthquakes can also be measured by such GPS geodetic methods. A remarkable development is the mapping of earthquake displacements (to about 3 cm) over areas of hundreds of square kilometres around the epicentre by satellite radar imaging, comparing images before and after the event (Massonet *et al.*, 1993).

A good example of plate motion is that at the southern Alaskan coast where the Pacific plate is thrusting north-westwards underneath Alaska, whereas along the nearby coast of North America strike-slip (horizontal) motion is dominant (Figures 4.2 and 5.4). Carbon dating along the Alaskan coast has shown that it had been undergoing submergence for some 1000 years before the 1964 M_s 8.4 earthquake. The earthquake produced uplift over a wide area as the plate rebounded upwards (Figure 5.5), as well as thrust to the north-west (Plafker, 1969, 1972). Similar pre-earthquake submergence and seismic uplift on a large scale seems to have occurred about 300 years ago along the coasts of Washington and Oregon states (Atwater *et al.*, 1991). The Pacific plate is also subducting beneath Japan and the Eurasian plate from the east and the Philippine plate from the south-east, producing the high seismicity and many volcanoes in Japan.

The most famous seismic zone is the region surrounding the San Andreas Fault system in California (Figures 5.6 and 5.7), which is a transform fault system between the Pacific plate and the North American plate. The San Andreas Fault is part of the fault system (or series of faults) which stretches for 1100 km or more from the East

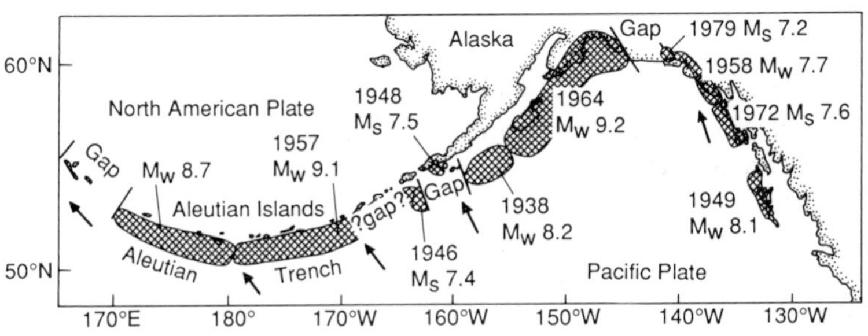

Figure 5.4 Ruptures zones (hatched areas) of large shallow earthquakes from 1930 to 1979 and three large seismic gaps along the plate boundary in the Alaska–Aleutian area. Arrows show the relative plate motion. (After Sykes, 1983)

Figure 5.5 The Hanning Bay Fault, about 4 m vertical displacement, in the 1964 Alaskan earthquake. Tectonic deformation associated with the 1964 Alaska earthquake. (Photo by George Plafker, US Geological Survey; from *Science*, **148**, 1657–1687. Copyright 1965 by the AAAS)

Pacific spreading ridge in the Gulf of California off Mexico to the Juan de Fuca ridge off northern California, taking up the relative motion of the Pacific and North American plates (Powell and Weldon, 1992) (Figure 5.6). It is 80–150 km wide and largely within the continent.

The San Andreas Fault is near-vertical and of strike-slip nature (horizontal relative motion), with some north-east–south-west directed compression which has produced both mountain and valley topography. In the last 30 million years dextral (right-lateral) movement has offset correlated rocks by up to 320 km (Wallace, 1990; Brown *et al.*, 1992). Average slip rates for the late Quaternary have been as much as 24 mm/year. Strike-slip displacement can be clearly seen in the Carrizo plain desert area (Figure 5.8) where the low erosion rates have been less than the movements. Seismicity of the system is generally in the upper 12–15 km of the crust. (Stover and Coffman (1993) list US earthquakes).

Another important seismic zone is the 'Alpide' seismic belt which extends from Indonesia to Burma, to north India and southern former USSR (Figure 1.9), through Iran and Turkey with its 1500 km North Anatolian Fault zone (Figure 5.9), then through Greece and Italy to North Africa. The North Anatolian Fault (Barka, 1992) is a right-lateral strike-slip fault like the San Andreas, but is even more active and has suffered 18 disastrous earthquakes in the past thousand years, including those near Erzincan (Turkey) in 1939 and 1992. The Alpide Belt is important because of the high population and poor housing causing many of the deaths from earthquakes (e.g. the Armenian disaster of 1988 and the Erzincan events). Another important fault is the Dead Sea Fault, an approximately 1000 km long left-lateral strike-slip structure with a history of seismicity (Arieh, 1994).

The seismicity of the Alpide Belt is related to the northward drift of

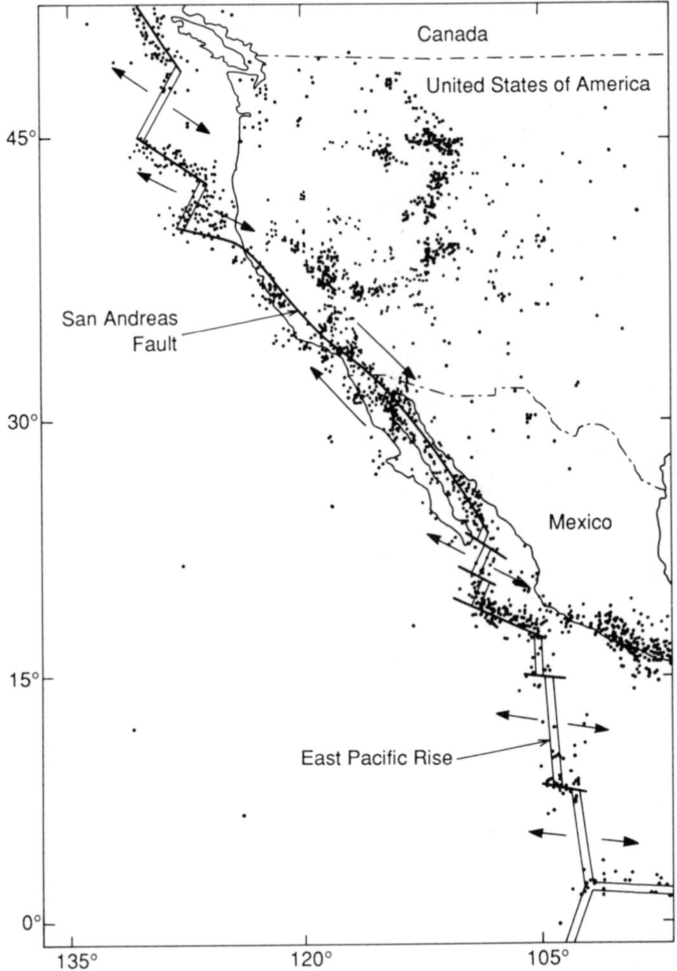

Figure 5.6 The San Andreas Fault as a large-scale transform fault. The double lines show the spreading ridges and dots represent epicentres. (After Wallace, 1990 and USGS)

Africa, Arabia and India and the westward motion of Anatolia (Turkey) (Oral *et al.*, 1995; Ambraseys, 1992). The high seismicity of the Himalayas and of China is also related to the Indian push to the north at and against Asia at about 2 cm/year. This has produced the Himalayas (now rising at about 1 cm/year) and thrusts under Tibet and has raised this highest and largest plateau on Earth as much as 5 km, and giving it a thick crust (Ni and Barazangi, 1984; Molnar and Lyon-Caen, 1989). The great Assam earthquake of 1897 caused an uplift at one point of 10.7 m. Recent deep reflection surveys have revealed a north-dipping reflector beneath southern Tibet probably marking the thrust fault along which the Indian plate is underthrusting southern Tibet (Zhao *et al.*, 1993).

This has also produced combinations of strike-slip and normal faulting in China (Zhou *et al.*, 1983); for example, the 1600 km long Altan Tagh Fault, a major strike-slip fault which bounds the Tibetan

Figure 5.7 The San Andreas Fault in the Mecca Hills, Coachella Valley, California. (Courtesy R. E. Wallace, USGS, Menlo Park, CA)

Figure 5.8 A pair of streams offset by the right-lateral slip of the San Andreas Fault (left to right). (Courtesy R. E. Wallace, USGS, Menlo Park, CA)

plateau and is very active. Also there is the 3000 km long Tancheng-Lujiang Fault system, again strike-slip. The Chinese have records of earthquakes going back to 780 BC, at least for many larger shocks. A magnitude 8 (estimated) event in 1556 caused around 200 000 deaths in Shensi province where many people lived in caves in soft loess (thick wind-blown dust and sand deposits), plus another 630 000 who,

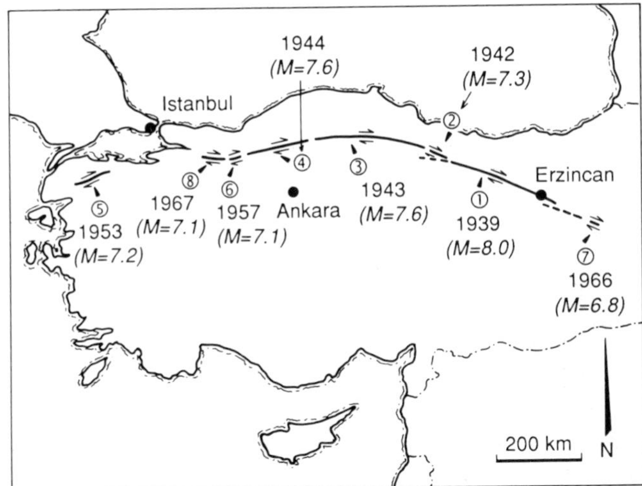

Figure 5.9 Fault displacements along the North Anatolian Fault associated with large earthquakes (1939–1967). (After Allen, 1969)

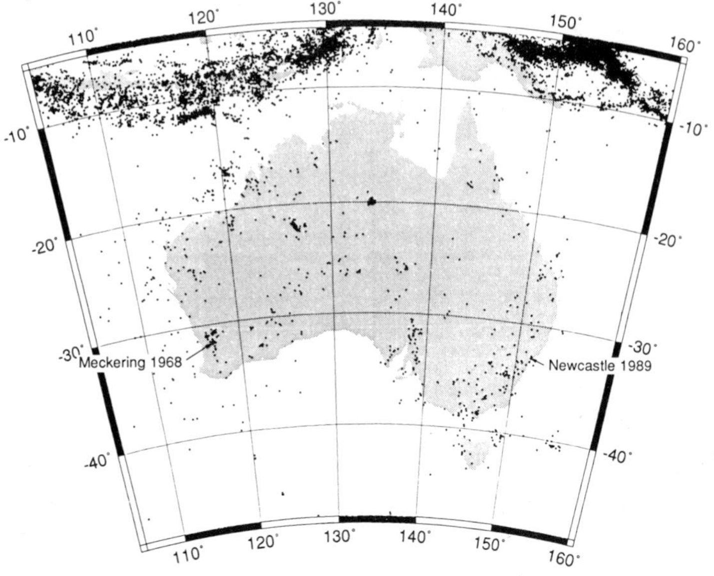

Figure 5.10 Seismicity of Australia and parts of New Guinea and Indonesia, showing epicentres of earthquakes >3.9, 1900–1991. (Courtesy Director, Aust. Geol. Surv. Org.)

it is reported, died later from starvation. In 1976 the M7.8 earthquake at Tangshan, 160 km east of Beijing, caused over 242 000 deaths and 700 000 injured (Chen Yong *et al.*, 1988), despite efforts to predict such events (p. 80). Over long periods the average death rate from earthquakes worldwide is about 15 000 per year and 17 000 since 1900 (Lomnitz, 1994).

The Indonesian arc is another active region, having 129 active volcanoes as well as earthquakes. The outer edge of the Australian

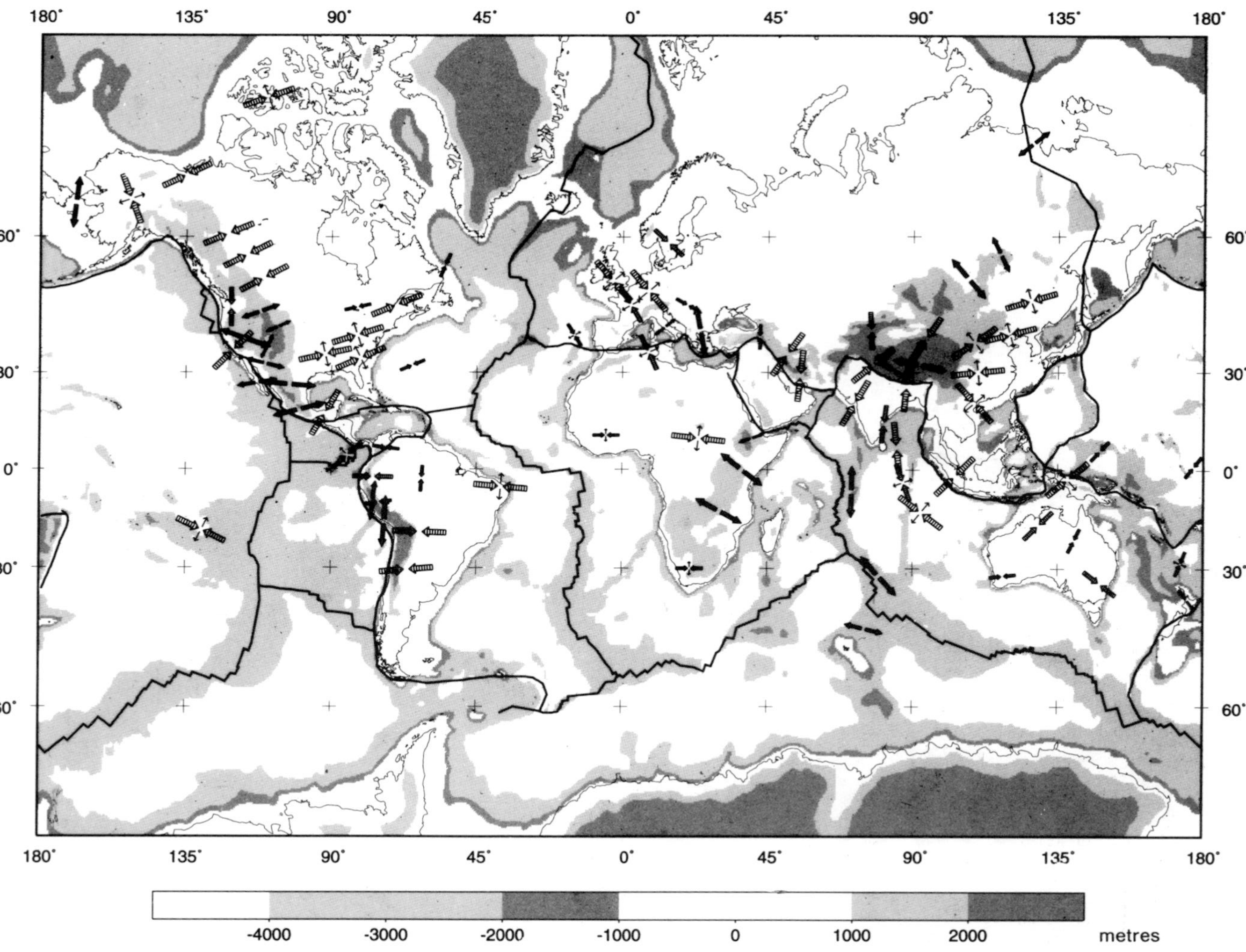

Figure 5.11 A generalized global stress map. Large arrows show mean stress directions, stress regime is indicated by different arrow sets (inward arrows only = thrust regime, big inward and small outward = strike-slip regime, outward only = normal faulting). (Courtesy of M. L. Zoback, 1992, US Geol. Surv.)

continent is being subducted northward beneath the eastern Sunda Arc (McCaffrey and Nabelek, 1984; Figure 5.10). The great Sumba earthquake (M7.9) of 1977 was a large normal faulting movement which supports the idea that slab-pull downwards is an important force in plate movement.

Although most earthquakes occur at plate boundaries, many damaging shocks have occurred within plates (intraplate), e.g. in eastern USA and Australia and, as we saw, China, which is made up of a number of accreted blocks, has one of the highest rates of intraplate activity. Intraplate earthquakes are often difficult to relate to surface faults, so that their origin is unclear and their study more difficult (Sykes, 1978; Long, 1988; Johnston and Kanter, 1990). According to Basham and Gregerson (1989), most earthquakes above M6 in stable continental interiors are associated with old continental rifts or passive margins.

In most intraplate regions, especially shields and old ocean basins, upper mantle temperatures are low and the upper mantle strong. However intraplate regions such as in North America, the Australian continent and Fennoscandia (Gregerson *et al.*, 1991) are under horizontal stress, as apparently are all the continents (Figure 5.11). Mid-plate stresses are believed to be largely from forces at plate boundaries, largely ridge push and continental collision (Whitmarsh *et al.*, 1991; Zoback, 1992).

Antarctica appears to be almost aseismic as no earthquakes larger than M5 have been located in that continent. This may be partly explained by the load of glacier ice, averaging about 2 km thick. Australia is fortunate in having a comparatively low seismicity, thanks to being within a plate and containing no plate boundaries, like Antarctica, whereas New Zealand straddles the eastern margin of the Australian plate (Figure 1.9) and includes the 650 km Alpine Fault (Anderson and Webb, 1994). Nevertheless, about 500 earthquakes are located each year in Australia, two small towns have been wrecked in Western Australia (Gordon and Lewis, 1981) and 13 people died in the M5.6 Newcastle earthquake of 1989 (McCue *et al.*, 1990). This was the first case of multiple deaths in Australia from earthquakes, thanks to the low seismicity and low population density over the greater part of the country and a good standard of housing. It is not so much the earthquake which kills but the structures people live and work in.

Apart from natural (tectonic) earthquakes, events also occur as products of volcanism (see p. 87), deep mining (Gibowicz, 1990; Fairhurst, 1990; Young, 1993), quarry blasts and possibly large dams (Meade, 1991).

Chapter 6
Earthquake mechanism

6.1 EARTHQUAKE FRACTURE

An earthquake is a rupture within the Earth caused by stress. In 1891 the Mino-Owari (Japan) earthquake revealed clear evidence for the faulting origin of earthquakes with a 110 km long fault and 6 m vertical movement (Howell, 1990). However, faulting as the main cause of earthquakes was only accepted gradually, partly because faults only occasionally break the surface. The famous San Francisco earthquake of 1906, with its surface faulting and the geodetic surveys carried out, confirmed this and provided the evidence for the first modern model of earthquake mechanism. Reid (1911) used the survey data to show that the great energy of earthquakes is stored in the rock by the gradual build-up of strain over time, often over hundreds or even thousands of years (Figure 4.3). This is known as Reid's Elastic Rebound Theory, and has been verified in general by modern research, at least for shallow events (see p. 68 on deep earthquakes). Earthquakes are often modelled by slider blocks on a friction base with stress applied via springs (Figure 6.1); two-dimensional models may be more realistic.

In the San Francisco earthquake the western (Pacific plate) side of northern California moved up to 6 m north relative to the eastern (North American plate) side of the strike-slip San Andreas Fault over a length of 425 km (Brown *et al.*, 1992) (Figure 6.2). This motion was in the same (dextral) direction as the crust had been strained before 1906, as shown by the geodetic data. This straining is continuing and will no doubt produce another large earthquake eventually, next time possibly further south where the 1857 Fort Tejon earthquake (Figure 6.2) took place (Wesson and Wallace, 1985).

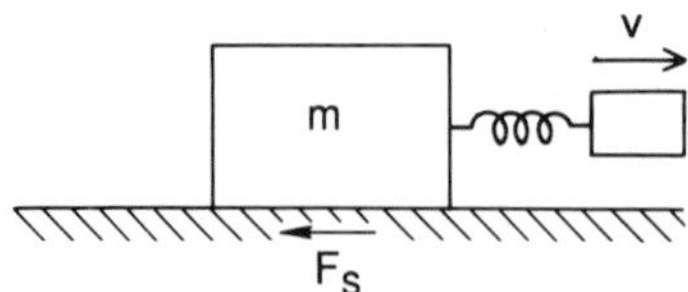

Figure 6.1 A simple slider model of earthquake mechanism. v = velocity, m = mass and F_s = static friction

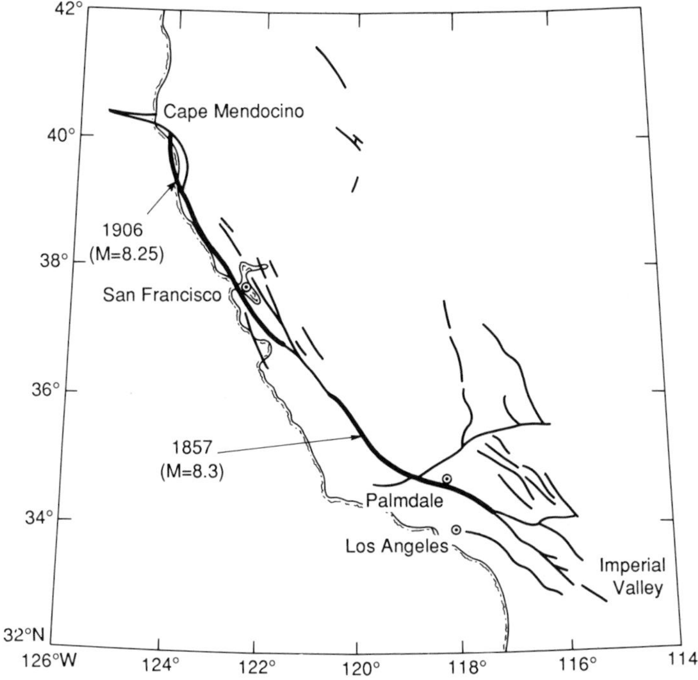

Figure 6.2 The sections of the San Andreas Fault displaced in the 1906 and 1857 great earthquakes (thick lines) (USGS)

Measurements show that the continents are largely in a state of strong horizontal stress. Most of the stress in the continents on both sides of the North Atlantic is caused by ridge push from the cooling and thickening of the lithospheric plate away from the Mid-Atlantic ridge (Whitmarsh *et al.*, 1991). In Fennoscandia, isostatic uplift of the crust with the melting of the glaciers may also be a factor. A cubic metre of rock in the upper seismogenic crust can store about 100 joules of strain energy, and as there are perhaps 10^{15} m^3 of strained rock around a large fault, there is plenty of energy to release, all in a few minutes or seconds, though it may have taken centuries to accumulate (e.g. see Plafker (1969) and Savage *et al.* (1986) for the Alaskan event of 1964).

After a large earthquake, aftershocks (p. 69), surface slips and slow fault creep may occur for a year or more. It has been believed that fracture occurs when the stress increases to beyond the strength of the brittle lithospheric rock, but surprisingly, stress measurements adjacent to the San Andreas Fault indicated very low stress levels (Healy and Zoback, 1988), perhaps because of high fluid pore pressure. A borehole has been drilled at the Cajon Pass 4 km from the fault to a depth of 3.5 km without finding any right-lateral shear stress or frictional heat (Zoback and Lachenbruch, 1992). Thus the San Andreas Fault seems quite weak, although 3.5 km is shallow compared to the 15 km maximum depths of the foci.

In a series of papers, Sibson (1989) and others have proposed that rising fluid pore pressure is a mechanism for brittle failure. Faults and shears localize fluid migration, which in turn affects fault strength (fluid migration in active faults is also important in the genesis of ore bodies). Chester *et al.* (1993) discuss the structure and weakening mechanisms of the San Andreas Fault. *In situ* stress measurements in many active *intra*plate areas do not show such weakening. Stress drops produced by intraplate earthquakes are on average six times higher than for interplate events, implying higher frictional strengths according to Scholz *et al.* (1986). Understanding exactly how earthquakes work will depend on learning how strong rock bonding is weakened.

Reactivation of old faults is probably more common than new fracturing. When fracture occurs across an old fault its strength will be governed by the frictional force across the fault. This increases with pressure and thus with depth, but is decreased by fluid pore-pressure which reduces the effective pressure across the fault and so the friction. Also, faults contain soft gouge produced by past movements and it has been suggested that gouge plays an important part in the mechanics of some fault motion (Scholz, 1990).

At greater depths, with the increasing pressure and temperature, rocks pass through the brittle–ductile transition and undergo plastic flow instead of brittle failure (Sibson, 1982: Scholz, 1990; Kanamori, 1994). There is a lot of discussion on the mechanism of earthquakes at deep levels, 100 to almost 700 km, as at these depths ductile flow would presumably remove the shear stresses (Frohlich, 1994a; Green, 1994). Deep earthquakes occur in subducting slabs in which temperatures may be as much as 1000 °C colder than the surrounding mantle. For depths down to 300 km in subduction zones, brittle failure is believed by some to be made possible by water subducted along with the lithosphere or released by hydrated minerals (Meade and Jeanloz, 1991; Green, 1994). For very deep colder subduction zones where foci occur to almost 680 km, metastable olivine breaks down suddenly into two very dense phases with anti-crack faulting, according to Green (1994).

It is found in laboratory measurements that at a stress of about half the strength of the rock, the rock begins to dilate because of the formation of many small cracks (Kasahara, 1981). However there is little evidence for significant dilatancy before earthquakes (Turcotte, 1991). Figure 6.3 shows a possible varying stress pattern, the stress drop produced by the faulting and the sliding friction during the movement.

Recorded waveforms are now commonly mathematically inverted to produce models of the slip on the faults (Dziewonski *et al.*, 1987). This has shown that most earthquakes have quite complex sources, some are multiple events, and simple ruptures are the exception

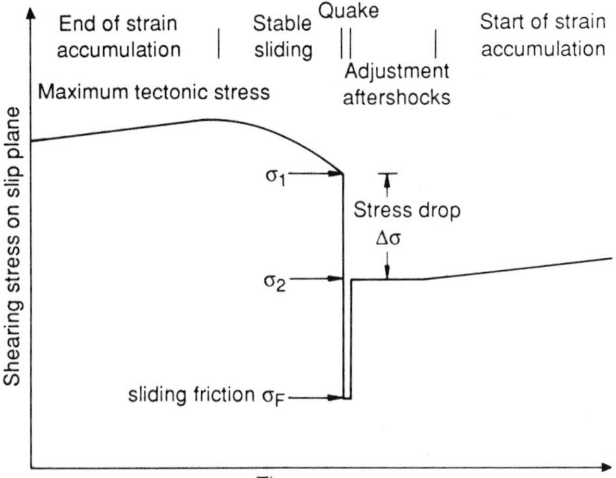

Figure 6.3 A hypothetical stress history of an earthquake. The scales are probably non-linear. (After Fitch, 1979)

(Brune, 1991). Major thrust faults produce most of the moderate earthquakes and many large ones. Earthquakes are the result of ripping along the faults at varying speeds of kms per second, radiating seismic waves. The speed variations are because of changes of strength, particularly at asperities (strong points), along the faults (Brune, 1991). Such complexities on the faults produce the higher seismic frequencies while most of the long-period radiation comes from weak areas. Energy is dissipated as seismic waves and to a smaller extent as heat of friction on the faults.

There appears to be evidence also of 'slow' earthquakes, in which the rupture moves slowly over tens of minutes, particularly on oceanic ridges, and even 'silent' events. Slow earthquakes are difficult to detect on seismographs, although Beroza and Jordan (1990) have reported that they can detect Earth oscillations from them on accelerometers.

Amongst the phenomena reported with earthquakes are sounds, presumably produced largely by P waves entering the air and the body; visible waves in the ground; and the so-called earthquake lights which have been variously explained as being from voltages produced by fracture, plasmas, electron excitation (Brady and Rowell, 1986), piezo-electric effects from the high rock stresses, and cavitation (Johnston, 1991).

6.2 AFTERSHOCKS

After a moderate to large earthquake there are hundreds or even thousands of aftershocks, with hundreds per day occurring after a large event (Figure 6.4). The largest aftershock is usually at least one magnitude lower than the main shock, and the others mostly much

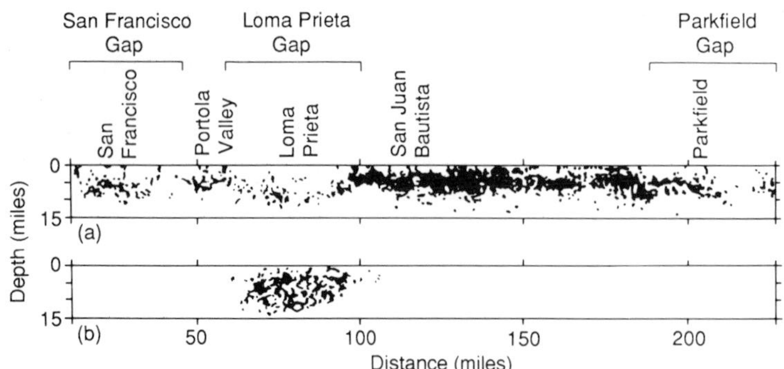

Figure 6.4 Cross-sections showing the seismicity along the San Andreas Fault from north of San Francisco to south of Parkfield for 1960–1979. (a) The dense zone of activity south of San Juan Bautista in the creeping section of the San Andreas Fault. The Loma Prieta section has shown little activity in the previous 20 years. (b) The formerly quiet Loma Prieta zone was filled by the main shock and aftershocks of the Loma Prieta event (1989). Note also the Parkfield gap to the south. (After USGS Yearbook, 1990)

smaller still. Their frequency decreases sharply with time after the main shock, but there may still be isolated aftershocks a year or two later. The aftershock annual frequency n follows Omori's relation, $n \propto 1/t$ (t is the time after the main shock), which Utsu later generalized to $n \propto 1/(t + t_0)^p$ where t_0 is a constant and p usually near 1. Foreshocks may follow a similar pattern where t is the time before the main shock.

Aftershocks are presumably caused by the strain not being fully released in the main event, and by a rearrangement of stresses in the region. The aftershock pattern indicates the position of the fault or even of a moved block. The aftershocks usually occur within two source-lengths (the main shock fault-length) of the main epicentre. It is claimed that irregular ruptures may produce many aftershocks whereas a 'clean' linear rupture will produce few.

In the case of the Loma Prieta (California) earthquake of 1989 most of the aftershocks were confined to the fault plane, with the main shock at the bottom (Figure 6.4). Two aftershocks of magnitude 5.0 or larger were recorded and 48 of 4.0 or larger in a zone 66 km long (McNutt and Sydnor, 1990).

6.3 FOCAL MECHANISMS

To conserve angular momentum earthquake sources must (usually) consist, in effect, of two opposing couples, i.e. a double couple: the stress couple (Figure 4.3) and the reaction of the surrounding material to it. Mathematical modelling has produced useful predictions of the patterns of wave amplitudes at distant stations. Thus seismograms not only record the arrival time and amplitudes of earthquake waves or 'phases', thus enabling estimates of focal positions, depths and magnitudes, but also the type and direction of fault motion if there are enough clear records from various observatories (Figure 6.5). This is known as focal mechanism or direction of motion study (Herrmann, 1975; Das *et al.*, 1986), pioneered by Nakano in 1923. Compressions and rarefactions are produced by fault motion; which one the station

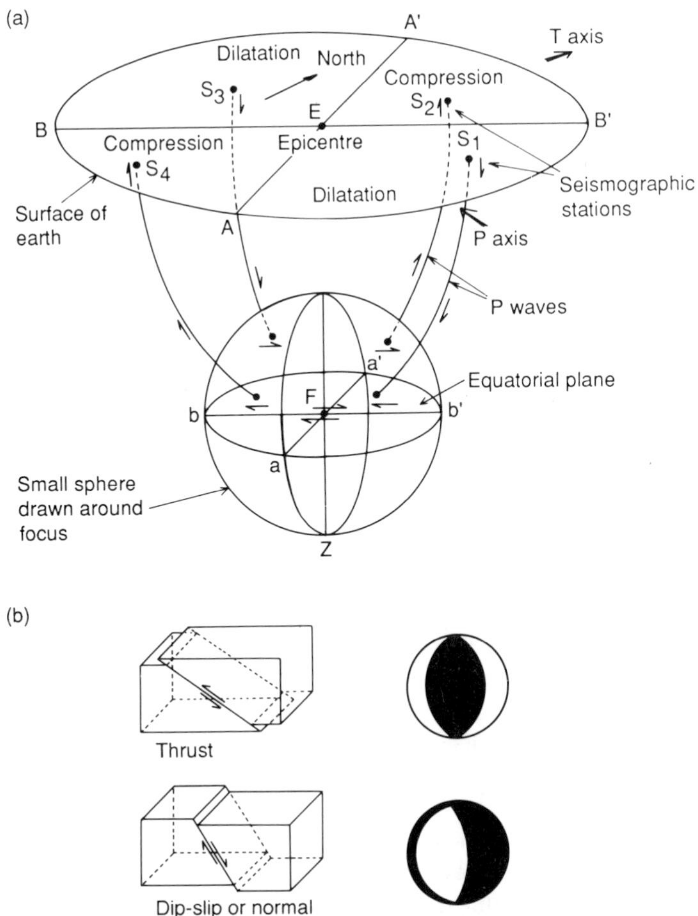

Figure 6.5 (a) The patterns of first motion recorded at stations around the epicentre, plotted on a sphere around the focus F by projection. (b) The three main types of focal mechanism (thrust, normal and strike-slip) and their projections on the lower hemisphere (black for compression). (After Bolt, 1993; Mooney, 1989)

records depends on the direction from the fault of the station. However, explosions radiate compressions in all directions (plus possibly some release of local stress). This assists in the recognition of nuclear explosions. There are also some other non-double-couple cases, e.g. in volcanic and geothermal areas (Frohlich, 1994b).

The vertical component determines whether the first P motion is a compression or a rarefaction. If the ground moves up, which should be up on the record, it shows that the P arrival is a compression pushing away from the focus. This is because rays always refract down into the Earth and back up to the recording station. If the first motion is down, the motion is a rarefaction pulling back towards the focus.

Modern digitally recorded data also allow the use of waveform modelling to estimate focal mechanisms.

By plotting the various compressions and rarefactions from many stations on a map, the movement at the fault can be estimated. For a local earthquake a plane map can be used, but for a distant epicentre a stereographic projection is used. The data are displayed on the equatorial plane of the focal sphere (a small sphere around the focus, Figure 6.5) by extrapolating back along the ray. The quadrantal (four section) distribution of compressions and rarefactions is diagnostic of the kind of faulting, e.g. the strike-slip, reverse and normal fault patterns. The fault direction is given by one of the great circles dividing the pattern; which one is decided from surface faulting, from past events, aftershocks or local structure. If the motion is not pure strike-slip but includes a component of dip-slip, the circle representing the plane perpendicular to the fault will be tilted towards the circumference.

Focal mechanism data can also indicate the directions of the principal stress axes (Figure 5.11), taken to be at 45° to the fault directions, although possibly biassed by old weak faults and local conditions. An interesting phenomenon is the Doppler effect showing up in surface wave data, i.e. the increased frequency of waves at stations in the direction the fault is moving and vice versa.

Fault plane studies show the directions of movement at plate boundaries and have been very important in confirming plate tectonic models. For example, faulting along mid-ocean ridges is either normal or strike-slip, depending on whether the epicentre is on the ridge itself or on a transform fault. At subduction zones focal mechanism studies indicate that where the plates begin to bend, the earthquakes are the result of the bending. At great depths they are the result of compression as resistance to penetration increases. In the Loma Prieta (California) 1989 earthquake a right-lateral (SE–NW) displacement of 1.6 m and reverse faulting up and to the east of 1.2 m was inferred. The Santa Cruz mountains were uplifted about 36 cm, but there was no convincing evidence of primary surface faulting (McNutt and Sydnor, 1990). The presence of marine terraces in the area is evidence of similar uplifts from past earthquakes.

Chapter 7
Earthquake engineering

The death rate from earthquakes worldwide is about 17 000 per year since 1900 (Lomnitz, 1994). But earthquakes do not kill so many as do the affected buildings, particularly masonry buildings, rubble stone masonry being the worst. In a strong earthquake a structure may accelerate as much as several times the maximum ground acceleration, depending on the building's mass, stiffness and damping. Thus earthquake building codes have been introduced in many countries, at least in their seismically active zones (Coburn and Spence, 1992). Unfortunately, codes may only affect new buildings. The frequency of fatal earthquakes on land is increasing as populations increase and spread. Unfortunately some of the highest population growths have been in seismically or volcanically active areas such as the Philippines, Indonesia and Mexico.

Because inertial forces are involved, light and supple structures such as timber or fibro housing are safer than are those of brittle and heavy brick or stone. Old stone buildings built without or before the introduction of building seismic codes are particularly at risk (e.g. the M6.4 Latur (India) earthquake of 1993 where poor buildings and foundations resulted in over 11 000 deaths). The 1994 Northbridge (California) earthquake caused over 60 deaths, whereas the Iran earthquake of 1993 of similar size caused 55 000 deaths because of poor structures with heavy roofs. Similar problems occurred in the 1995 Kobe, Japan and Sakhalin, Russia earthquakes.

Chimneys and tile roofs are also a danger to inhabitants. Brick masonry buildings may be made much safer with appropriate reinforced steel (Key, 1988; Juhasova, 1991; Krinitzsky *et al.*, 1993). Modern buildings with steel columns and beams and/or reinforced concrete perform quite well if on firm foundations, having flexibility and some ductility to absorb strong vibration. However, the Northridge (CA) event of 1994 caused surprising fractures in welds in steel frame buildings which may be more susceptible to strong shaking than previously expected (USGS Scientists and Southern California Earthquake Center, 1994). Experience shows that structures can survive major shaking if designed for the horizontal forces recommended in the codes and there is solid ground beneath them. A particular concern is that of nuclear power stations and dams and their safety in strong ground movement.

The most powerful vibrations from an earthquake are in the frequency range 0.5–5 Hz (Joyce, 1991) and at near and regional distances the strongest ground motions are produced by the S and Lg waves (Boatwright and Choy, 1992). A typical building of 10 storeys has a natural period of about 1 s. Each storey adds about 0.1 s and a 20-storey building has a period near 2 s. Taller buildings have the advantage of flexing more than short stiff buildings, and are usually designed to bend with the wind. Thus those over 20 storeys may fare relatively well. But with their longer periods they are sensitive to distant earthquakes, as at Mexico City in 1985. People in the upper storeys of tall buildings in Perth, Western Australia, have been frightened by earthquakes in Indonesia 2400 km away.

Anti-seismic design may include estimates of the buildings' responses and the frequencies of the various vibration modes using computer models, or the use of models of the buildings on large shake-tables (Figure 7.1). The design would include damping and ductility in the structure to absorb vibration energy.

One source of damage is the slapping of one building by another, particularly if they are of different height and therefore different natural frequency. Some modern designs include rubber bearings (base isolation) for buildings to increase absorption of ground motion and one new Tokyo building is to have moving counterweights reacting instantly to counter the effects of strong vibration detected by sensors. In the Kobe (Japan) earthquake of 1995, damage was severe partly because the fault rupture was towards the city, whereas

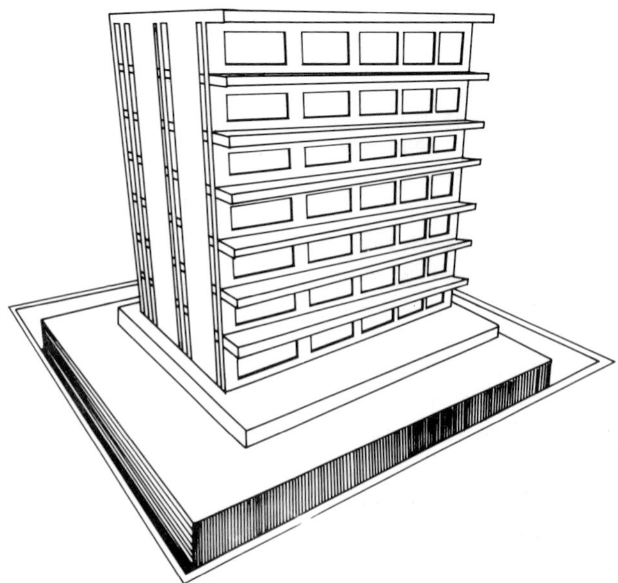

Figure 7.1 The large shaking table of the Japanese Ministry of Construction. It can support a load of 17 t (e.g. a replica of a building) and shake it with a vertical acceleration of 1g and horizontal acceleration of 2g with periods from about 0.2 to 0.9 s. (After Dr T. Hisada, Building Res. Inst. Japan)

in the Northridge event the focus was under the suburbs but ruptured away from the city.

The problem of the nature of strong ground motion from earthquakes has been reviewed by Boore (1977), Heaton and Hartzell (1988) and Anderson (1991). Amplitudes can be several times higher on ridges and mountain tops. Measurements of ground acceleration and velocity are important as they can be related to the forces on buildings (e.g. see Jacob and Turkstra, 1989). Accelerometers are a special type of seismograph designed to have a response proportional to the acceleration and have a low gain so that they do not saturate during an earthquake. It is now common practice to set up networks of accelerometers in seismically active areas to record ground and building accelerations when the instruments are triggered by a tremor and so assist anti-seismic building design (Trifunac and Brady, 1975). The US Geological Survey currently operates more than 1000 such instruments (Maley *et al.*, 1990) and the California Division of Mines and Geology more than 450.

Recent recordings such as at Northridge show ground forces near faults even higher than predicted. Short-duration accelerations as high as $1g$ or even $2g$ ($g = 9.8\,m/s^2$, the acceleration of gravity) have been recorded by accelerometers in some earthquakes, but this is not as important as the longer time values. Amplitudes at a certain distance (Figure 7.2) from an epicentre vary a great deal depending on geological conditions. Ground vibrations are magnified several times in areas of soft soils, young sediments and artificial fill, particularly if waterlogged. This is caused by the low rigidity (thus low S velocity) of the material and resonance between the seismic waves and the sediments, soil and fill of similar natural periods of oscillation. Many cities around the world (e.g. Tokyo, Mexico City and San Francisco) are built partly on such materials, being adjacent to water. The duration of shaking is also a very important factor in building damage and it also increases on softer foundations.

The Loma Prieta, California, magnitude (M_s) 7.1 earthquake of 1989 was the largest to shake San Francisco since the great 1906 event (US Geological Survey Staff, 1990; McNutt and Sydnor, 1990; Hanks and Krawinkler, 1991). Strong shaking was associated with recorded accelerations of over $0.1g$ from the S and surface waves. Much significant damage was far from the epicentre in San Francisco, about 100 km to the north (McNally and Ward, 1990), showing the importance of the structures involved and of the geology and foundations (Figures 7.3 and 7.4). Fatalities in San Francisco and Santa Cruz (near the epicentre) were largely in unreinforced masonry buildings, causing over 60 deaths and direct losses of about $6 billion. Building and overpass collapses in San Francisco were produced where the structures were sited on bay fill, despite the epicentral distance. The Marina district of San Francisco was even built on the rubble of the 1906 earthquake. At Oakland, where the viaduct

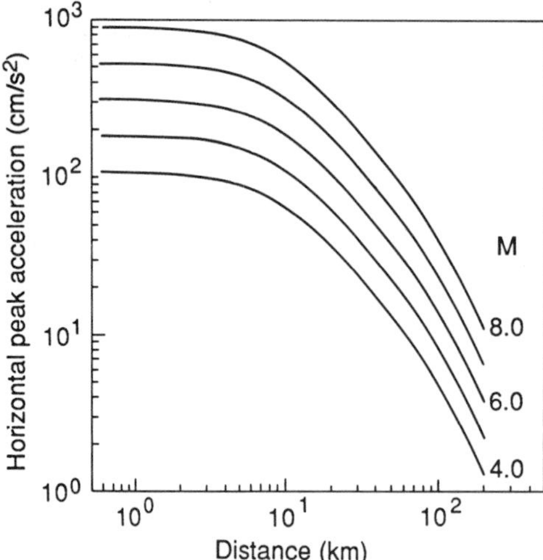

Figure 7.2 Mean ground motion attenuation curves of peak horizontal acceleration versus distance and magnitude. (After Joyner and Boore, 1981)

Figure 7.3 The Cyprus viaduct, Oakland, after collapsing during the Loma Prieta, California, earthquake of 1989. (Photo by Howard Wilshire, courtesy USGS, Menlo Park, CA)

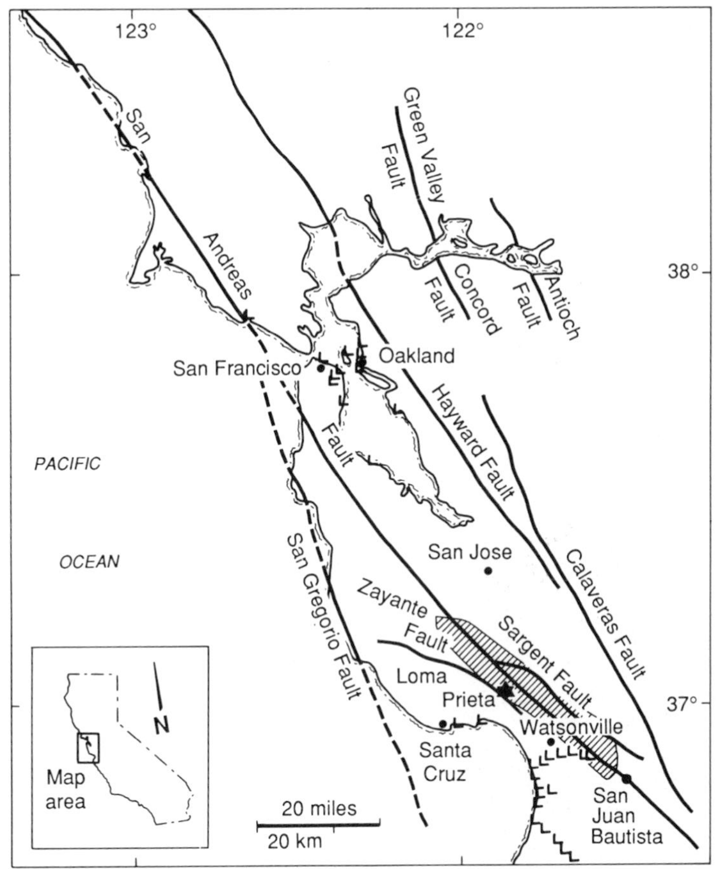

Figure 7.4 The major faults in the San Francisco area and the location of the Loma Prieta earthquake. (USGS Yearbook, 1990 p. 3)

Figure 7.5 Earthquake induced liquefaction and landslides damaged many houses in the Turnagain area of Anchorage, Alaska, in the 1964 earthquake. (Photo by Clarence Allen, courtesy of the California Inst. of Technology)

Figure 7.6 Aerial photograph of the Huascaran avalanche that destroyed the towns of Yungay and Ranrahirca, Peru, in the 1970 earthquake. (Courtesy of George Plafker, US Geol. Surv., and of the Seismological Society of America)

collapsed (Figure 7.3), accelerations on soft soil were about three times those at Berkeley (on hard rock), and durations were five times as great (Bolt, 1991). A broken gas main and water mains resulted in a serious fire in the Marina district. Similar situations occurred at Kobe, Japan, in 1995.

At Mexico City in 1985 resonances in ancient lake sediments under parts of the city amplified ground motions from the Michoacan earthquake (M_s8.1), even though it was centred 370 km away (Anderson *et al.*, 1986; Singh *et al.*, 1988). It was followed by an M_s7.6 event. Soft clay under the city was 'tuned' to earthquake periods of 2 s, resulting in magnification of ground acceleration from about 4% of g to 20% of g (Joyce, 1991). Many buildings of 15–25 storeys have natural periods near 2 s. Much greater amplifications have been reported, particularly at 0.5 s (Singh *et al.*, 1988). In this disaster 412 buildings collapsed, including reinforced concrete structures on soft ground and over 3000 were badly damaged with $4 billion in property loss. At least 9500 people died and 250 000 made homeless. At Newcastle, Australia,

Figure 7.7 The fault scarp produced by the Meckering earthquake in Western Australia in 1968. The maximum uplift was about 2 m. (Photo by H. A. Doyle)

which is built on sands and muds of old river channels, the 1989 M5.6 event caused $1.5 billion worth of damage.

In epicentral areas the strong vibrations can lead to quasi-liquefaction of the soil, particularly with loose saturated sands. Then soils will flow downhill. In 1964, this occurred at Anchorage, Alaska where homes built on glacial moraine were carried downhill in a landslide following the liquefaction of the moraine during an earthquake (Figure 7.5). In the M7.5 shock at Niigata, Japan, in 1964, one-third of the city subsided by up to 2 m due to the vibrations compacting the soil. Water was expelled from the ground by the compaction and flooded the streets, and some buildings tilted over. Avalanches may occur with the rise of liquid pore pressure and loss of adhesion. In the M_s7.7 offshore Peruvian earthquake of 1970, the steep slopes of the Andean mountains induced an avalanche of great speed and volume which travelled 14 km, burying whole towns (Figure 7.6). Cities near seismic zones should be mapped in detail to show areas of thick and/or wet soil and fill to aid determination of anti-seismic building codes. Site response can be checked by recording small tremors and microseisms (Field *et al.*, 1990).

As well as vibration effects, there is the possibility in larger shallow shocks of surface faulting (vertical and/or horizontal) directly damaging structures, e.g. roads, railways, pipelines and dams. The largest uplift so far recorded was 16 m in the Alaskan earthquake of 1899 and a maximum uplift of 11.3 m occurred in Alaska in 1964 (Figure 5.5 a separate fault). In the Precambrian Shield of Western Australia, at Meckering in 1968, vertical uplift of almost 2 m in a M6.9 event (Figure 7.7), plus a similar thrust to the west, produced a fault scarp 37 km long which twisted and telescoped railway and pipelines (Gordon and Lewis, 1980).

Chapter 8
Earthquake prediction

Much effort has been expended in the last few decades in the USA, Japan, Russia and China in the search for methods of earthquake prediction (Simpson and Richards, 1981; Rikitake, 1982; Turcotte, 1991; Lomnitz, 1994). Results so far have been disappointing (Wyss, 1991) and there is no certainty that reliable prediction is possible, although there have been a few forecasts (place but not time) and at least one successful prediction. A 'prediction' has been defined as a relatively precise statement giving the time and place of an event. Earthquakes are 'chaotic' phenomena and the timing can be affected by small changes in various factors. Also, faults of all sizes occur everywhere in the lithosphere with a fractal distribution. Earthquakes are also fractal in their magnitude/annual frequency relationship (p. 53).

The successful prediction was that of the M7.3 Haicheng earthquake of 1975 in China in which few lives were lost despite 90% of structures in the city being destroyed (Cao and Aki, 1983). Fortunately, the main shock was preceded by increasing foreshocks in the preceding hours and the populace evacuated homes and buildings just in time. In contrast, the M7.8 Tangshan (China) earthquake of 1976 was not preceded by any recognized foreshocks and was not predicted; it took some 242 000 lives (Chen Yong *et al.*, 1988). This was the worst earthquake in history since the Chinese disaster of 1556. About 25% of large earthquakes have a significant foreshock.

Seismic hazards are commonly investigated on probabilistic lines, this has been reported on in the analyses of the Panel on Seismic Hazard (1988) and by McGuire and Arabasz (1990) and Bolt (1991). For some areas statistical estimates of the probabilities of earthquakes of various magnitudes that may occur can be made and steps taken to strengthen or remove unsafe structures, but this is often financially and politically difficult. Berke and Beatly (1992) discuss various social aspects of planning for future earthquakes. There are great social and economic problems associated with wrong predictions.

Another possible development is 'real-time' warning in which fast computer epicentre locating by local networks may allow warnings to be issued before the comparatively slow seismic waves (particularly S and surface waves) arrive at distant cities. Mexico City plans to have a warning of impending shocks from earthquakes on the active western coast of Mexico, which is about 50 s away in seismic travel-time. Japanese railways use seismic data to warn of large events and automatically brake the 'bullet' trains.

In some fault zones it may be possible to measure the rate of fault movement (in earthquakes and fault creep) over historic and recent geologic time and so estimate the slip-rate (Thatcher, 1984). This might then be used to estimate the time required for the fault to accumulate enough strain for a certain magnitude earthquake to occur. Unfortunately, however, the amount of strain required to produce an earthquake is not known (Turcotte, 1991).

Some faults, e.g. a segment of the San Andreas Fault near Parkfield, California (Figure 6.4), appear to produce characteristic earthquakes with very similar epicentres, fault planes and perhaps return periods (Bakun and McEvilly, 1984; Roeloffs and Langbein, 1994). These Parkfield earthquakes have been magnitude 6 events with a possible average return period of 22 years, the last occurring in 1966. The first official prediction of an earthquake in the USA was made by the USGS in 1985 for a repeat of this 'characteristic' shock. The predicted time of the next Parkfield earthquake was January 1988 (± 5 years), so it is well overdue (Bakun and Lindh, 1985). However, the statistics used have been criticized and the fault's average behaviour may be misleading (see also Wyss, 1991; Lomnitz, 1994; Roeloffs and Langbein, 1994). The US Geological Survey (USGS) and the State of California have undertaken a densely instrumented study of the 30 km Parkfield section of the fault with more than 500 instruments using various techniques (US Geological Staff, 1990). This is the most heavily instrumented seismic region in the world.

Thatcher (1990) has concluded that earthquakes that re-rupture a segment of plate boundaries are not similar, and that characteristic earthquakes are the exception rather than the rule along the circum-Pacific plate boundaries. An important region at a plate boundary presently under threat is the Tokai area south of Tokyo where the Philippine plate is thrusting under Japan, and historic seismicity, including the disastrous 1923 Tokyo earthquake, point to danger. Other cities under threat include Istanbul, the capital of Turkey, near the western end of the North Anatolian Fault (Figure 5.9), and Wellington, the capital of New Zealand, through which the active Wellington Fault runs.

Seismic gaps are usually defined as a section of a plate boundary that has not ruptured for some expected recurrence time (e.g. the Alaska–Aleutian region; Figure 5.4) (Sykes, 1971; McNally, 1983). Two major subduction earthquakes off central Chile and Mexico in 1985 occurred in seismic gaps previously recognized, as did that which produced the tsunami on the Nicaraguan coast in 1992 (p. 90). Currently locked segments of the San Andreas Fault in California, which last broke in 1906 and 1857, may be seismic gaps (Figure 6.2; see also the Loma Prieta gap in Figure 6.4). However, the usefulness of seismic gaps has been denied by Kagan and Jackson (1995) who claim that places of recent activity have a larger than usual seismic hazard, whereas the 'gaps' around the Pacific are relatively quiet. Still, they may allow an estimate of the upper limit for the magnitudes that would occur in the gap.

The similar concept of seismic quiescence, a decrease in moderate seismicity several years before larger events, not necessarily on plate boundaries, has also been used as a warning by Wyss and Habermann (1988), who list 17 cases, and others. Wyss *et al.* (1992) report quiescence at Parkfield on the San Andreas Fault. This may also show as a change in the magnitude–frequency relation (*b* slope; W. D. Smith, 1986). The quiet period may be followed by an increase in activity before a larger event. However, Shaw *et al.* (1992) throw doubt on the quiescent model, claiming that long-term *increases* in activity are followed by large events.

One section of the San Andreas Fault between Parkfield and Loma Prieta is continually moving (creeping); together with micro-earthquake activity, this relieves most of the stress and makes a large event less likely. The creep motion is visible in the town of Hollister where sidewalks and foundations split from time to time. Fault creep also occurs with low seismicity on the Haywood and Calaveras faults east of San Francisco (Figure 7.4). The largest earthquakes are believed to be more likely to occur in the 'locked' sections of the San Andreas Fault where seismicity is low; one to the north where the famous 1906 earthquake occurred and one even more likely to the south towards the Mexican border (Figure 6.2), this section having suffered at least 10 large shocks in the last 1700 years. The last large shock was in 1857, the Fort Tejon earthquake, which ruptured over 300 km of the San Andreas Fault (Sieh *et al.*, 1989). The active San Jacinto Fault branches off to the west of the southern San Andreas Fault.

Until the last few years, activity has been remarkably low in comparison along these locked sections of the San Andreas Fault. However, in 1992 there was the nearby M7.5 Landers earthquake, 150 km east of Los Angeles in the Mojave Desert and the M6.6 aftershock 3 h later (see special issue of the *Bull. Seismol. Soc. Am.*, June 1994). Associated faulting broke the surface over a length of more than 70 km, the longest surface rupture in the US since 1906. These events seem to have triggered other events across western USA and may have increased the stress on the San Andreas Fault. Another possibility is the occurrence of large shocks along the eastern Sierra Nevada Range, the site of an earthquake in 1872 of about magnitude 8.

Allen Lindh of the USGS believes that the different characteristics along the San Andreas system are produced by the different rock types; the creeping section typically being a melange, perhaps with serpentine, which quickly becomes ductile with depth, while stronger rock along the southern section may remain stronger with depth.

Data suggest that the rate of slip of the San Andreas Fault and other faults averaged over millennia is the same as that at which strain has been accumulating across the faults over the past few years and decades (Argus and Gordon, 1990). Geodesists can now measure deformation to within several millimetres per year. Relative motion across the San Andreas and related faults as shown by such geodetic measurements is 3.5 cm/year of dextral strike-slip. This plus the 0.8 cm/year observed along the eastern Californian shear zone accounts for most of the observed relative motion between the North American and Pacific plates as measured by space-based geodetic methods such as the Global Positioning System (GPS), quasar interferometry and satellite ranging. The drift rate over geological time, indicated by magnetic anomalies in the Gulf of California, is 4.7 cm/year (DeMets *et al*., 1990; Lisowski *et al*., 1991). The remaining difference in strain rates is accounted for by movements in border areas. Also, in central California there is north-east directed horizontal compressive stress across the fault area which explains the uplift of the Coast Ranges and reverse faulting (Zoback and Zoback, 1989). The Southern California Earthquake Center at the University of Southern California, Los Angeles, is to set up GPS stations throughout southern California to monitor strain build-up there.

The Working Group on California Earthquake Propabilities (1995) estimate an 80–90% probability of an M7 earthquake or greater in southern California in the next 30 years. This poses a warning for the Los Angeles area, some 40 km away as do the buried thrust faults within the city area, such as the Whittier Fault which caused the damaging M5.9 earthquake in 1987 (the epicentre was 20 km east of the city centre). The more disastrous 1994 Northridge earthquake (M6.7, 33 dead, over 20 000 homeless and about $15 billion damage) occurred in an urban area and was the most costly American earthquake since 1906 (US Geological Survey Scientists and the Southern California Earthquake Center, 1994). It was also on a buried thrust fault (Davis and Namson, 1994), as was the San Fernando earthquake of 1971, M6.6 and 64 dead (Grantz, 1971). Both epicentres were in the San Fernando Valley.

A well-known danger area is the San Francisco area with the section of the San Andreas Fault to the west, which has not ruptured in a major event since 1868 but has creep and seismicity, and the Hayward Fault to the east (Figure 7.4). A revised probability estimate (made in 1990) for an M7 event or greater on the San Andreas or Hayward faults in the San Francisco Bay area in the next 30 years is 67%.

The portion of the San Andreas Fault along the southernmost portion of the 1906 rupture, and north of Santa Cruz, had been noted by Lindh (1983) and others of the USGS as having a greater than 47% probability of a major earthquake in the next 30 years. A 1988 USGS report described this segment as a dangerous seismic gap. The Loma

Prieta earthquake of 1989 (M_s7.1) occurred in the area (Figure 6.4). The Loma Prieta event, however, was not initiated on the main San Andreas Fault but on another fault several kilometres to the south-west and differed in having a reverse component (1.3 m upwards and 1.9 m to the north-west). There was a marked increase in seismicity in the 16 months before, otherwise no obvious short-term warning, but Gladwin *et al.* (1991) report a possible medium-term warning in strain measurements. Bolt (1991) has discussed this and other earthquakes in relation to the problem of estimating seismic risk.

The difficulty in detecting impending earthquake rupture is shown by the fact that monthly geodetic measurements of the distance from Loma Prieta and three other sites to a precision of about 3×10^{-7} indicated no significant changes before the Loma Prieta event. Also borehole strainmeters 10 km from the southern end of the rupture and 40 km from the epicentre recorded no strain events above noise level beforehand (US Geological Survey Staff, 1990). According to Takemoto (1991), it would be difficult to detect precursors using strainmeters and tiltmeters at distances more than twice the earthquake source dimension.

Further north, the western coast of North America is under threat from compression from the Juan de Fuca and Gorda plates (Figure 5.6), which are subducting under the North American plate in northern California, Oregon, Washington State and southern British Columbia (Savage and Lisowski, 1991). There is evidence of large subsidence in the past and buried trees, later uplifted in a large event, also some sand deposits interpreted as due to former tsunamis (Atwater *et al.*, 1991).

An important concept is that of 'asperities', rough or strong patches on faults which act as stick points where stress is built up and tend to govern the location of ruptures and when they occur (Aki, 1984; Sykes and Seeber, 1985). Such asperities may persist through many seismic cycles and produce characteristic events. The magnitudes of earthquakes are related to the lengths of the fault ruptures (Table 4.1); these segments often terminate at asperities, or fault offsets or bends which stop the rupture and may persist over many earthquake cycles (Sibson, 1986). Examples are the 'big bend' near Fort Tejon on the San Andreas Fault (Figure 6.2), which probably limited the big 1857 earthquake (about M8), and the offset and bend at the ends of the Parkfield section of the San Andreas Fault. The Loma Prieta event was also near a bend in the San Andreas Fault.

As we saw, the majority of earthquakes occur at plate boundaries and are associated with interplate movements (Isacks *et al.*, 1968; Minster and Jordan, 1978, 1987). So we understand the likely general locations of many future shocks, but not the timing, except under unusual circumstances such as at Haicheng, China (p. 80). Thus the big difficulty in prediction work is the time of any expected earthquake, and short-term prediction may never be very reliable, particularly if seismicity is a chaotic phenomenon. Usually a probabilistic approach is adopted rather than a deterministic one, calculating a mean repeat time and standard deviation, but usually the data are sparse and the statistics of the smaller and more frequent events have to be used as well. Ambraseys (1992) has argued that statistics based on short-term data alone do not provide a reliable estimate of seismic hazard and that clustering of major events makes simple statistical models inadequate. Study of historical and archaeological data as well is valuable.

Some progress has been made in long-term forecasting of larger magnitude shocks with the concepts of seismic gaps, seismic quiescence, slip-rate deficits and characteristic earthquakes. Short-term forecasting is difficult. Although increased seismicity is closely watched, as yet we have no way of distinguishing foreshocks of large events from small-scale earthquakes. About 25% of large events have foreshocks and Jones (1985) found that after any earthquake in southern California there is a 6% probability that a second earthquake of equal or larger magnitude will follow within five days or 10 km of the first. The 1934 and 1966 Parkfield events were preceded by M5.0 foreshocks.

Geological studies of earthquakes include mapping of surface faulting and secondary deformation from recent and past events (Allen, 1986; Vita-Finzi, 1986; Schwartz, 1987; Vittori *et al.*, 1991). As well as faults, folds, liquefaction features and sandblows are mapped. In the western USA, dating techniques for the late Quaternary have produced a large number of well-determined late Pleistocene–Holocene slip-rate averages, recurrent intervals and displacements per event for fault systems (palaeoseismology).

Slip rates from stream offsets, disturbed sediments and sandblows in saturated sands for example, are being used to estimate recurrence times on individual faults and sections of the San Andreas Fault and elsewhere. Sandblows form when strong shaking causes saturated sediments to 'liquefy' and be squeezed upwards through upper layers. Sieh *et al.* (1989) have dug trenches across a southern section of the fault to expose young sediment deposition, sandblows, buried scarps, etc., revealing at least 12 major earthquakes in the last 1700 years. Where organic matter is present radiocarbon dating has been used, with some error limits being less than 23 years. Many sandblows are seen in the New Madrid area of Missouri where the large earthquakes of 1811–1812 occurred. These earthquakes also dammed the Mississippi River creating a lake which still exists. Fault scarps may give rough evidence of their age by the amount of erosion and smoothing of the scarp.

Other possible indicators which have been measured in active areas include strain and tiltmeter variations (e.g. Savage *et al.*, 1986; Gladwin *et al.*, 1991), crustal stress (Zoback and Zoback, 1989), magnetic field variation (Johnston, 1989), radon and hydrogen outgassing from the ground (King, 1986), fault creep, elevation and gravity changes, water-well levels (favoured by Lomnitz, 1994) and various electric field values including resistivity and electromagnetic noise of frequency 0.1 Hz to 1 MHz (Tate and Daily, 1989; Park *et al.*,

1993; Varotsos and Kulhanek, 1993). The latter effects may be because of the motion of charges during rock microfracturing and some of the other possible indicators are likely to be associated with the formation of tiny cracks. The change in radon concentration before the 1978 Izu-Oshima-Kinkai earthquake may be a genuine precursor (Wakita *et al.*, 1988).

Some believe that slow or 'silent' earthquakes, in which smooth fault movement does not radiate high-frequency seismic waves, precede and probably trigger some normal earthquakes, particularly in oceanic lithosphere. Linde and Silver (1989) detected silent earthquakes with their volume strainmeters installed near faults and a slow precursor was detected before the great Macquarie Ridge event of 1989 (Ihmle *et al.*, 1993; Beroza and Jordan, 1990).

Prediction or forecast of *intra*plate earthquakes, including 'stable' continental crust, is particularly difficult as seismicity there is usually low, the epicentres tend to be scattered and associated fault zones are not necessarily identifiable (Long, 1988; Johnston and Kanter, 1990), e.g. in central and eastern USA (Nishenko and Bollinger, 1990) and in Australia (Doyle, 1984; Denham, 1988) (Figure 5.10). Intraplate seismicity tends to be patchy in time and space and the associated tectonics much less certain than for plate-edge ruptures. Deformation rates may be only one-thousandth that at plate boundaries, with return periods of perhaps thousands of years. However, major intraplate earthquakes may affect larger areas than do plate margin events. Nishenko and Bollinger (1990) state that central and eastern US has a probability about two-thirds that of California of having an earthquake causing similar damage. Magnitudes could be over 7 and depths shallow, causing serious damage, as occurred during the earthquake at New Madrid, Missouri, in 1811 (Jacob and Turkstra, 1989), where an ancient rift structure has been identified with unusually rapid strain accumulation (Liu *et al.*, 1992). Seismic velocities seem to be low in the upper crust of this and some other seismic areas. The Latur (India) M6.4 earthquake of 1993 caused widespread damage in 80 villages and over 11 000 deaths in an area with no historically reported significant events. It has been suggested that in stable continental interiors most earthquakes over M6 and all over M7 are associated with old continental rifts or old passive margins (Basham and Gregerson, 1989).

Finally it is worth quoting from Brune (1991):

> Although the likelihood of the successful use of short term precursor-based prediction of earthquakes has diminished, the ability to make long term (and perhaps intermediate term) estimates of earthquake probabilities, based on the concepts of characteristic earthquakes, seismic gaps, and long term slip constraints (in large part based on geologic evidence) has steadily improved.

Tsapanos and Burton (1991) have estimated the probable maximum magnitude in a period of 85 years for 50 of the most active countries of the world.

Chapter 9
Other seismic phenomena

9.1 VOLCANIC TREMORS

About 100 000 people have been killed by volcanic eruptions in the past century. Earthquakes are associated with volcanism, produced by the motion of magma, gaseous explosions and faulting (Decker, 1986; Latter, 1989; Tilling, 1989; Chester, 1993). They often occur as swarms of tremors (Figures 9.1 and 9.2), usually much smaller than the normal 'tectonic' earthquakes, but they are important indicators of impending eruptions, as at Mount St Helens, Washington State, in 1980 (Fehler, 1985; Weaver and Malone, 1987; Wright and Pierson, 1992) and Mount Pinatubo in the Philippines in 1991 (the latter was the largest eruption in over 50 years). Numerous short-period earthquakes (Figure 9.1) are believed to be due to brittle rock fracturing, and lower frequency tremors to gas and magma movements. Eruptions may be preceded and accompanied by a continuous vibration called 'harmonic' tremor because of its rather constant frequency. This is also an important warning and appears to be related to the presence and movement of magma. The recording of such events has been important in saving lives.

Most of the Earth's volcanoes are under the sea along the mid-oceanic ridges and this is where most (over 80%) of the Earth's volcanic activity occurs. However, those on land are of course of more concern and about 50 of the 600 known active volcanoes are in eruption each year according to Tilling (1989). It is a common practice to install a network of radio-telemetered seismographs around suspect volcanoes as an important part of warning systems (Malone *et al.*, 1981), e.g. at Pozzuoli in the volcanic area west of Naples in Italy, around various Japanese and Alaskan volcanoes, Rabaul harbour, New Guinea (which erupted in 1994, devastating the town), along some of the Cascade Range of volcanoes in the western USA (which include Mount St Helens and Mount Rainier), and at Long Valley, California (Figure 9.3). US and Philippine vulcanologists made a successful prediction of the 1991 Mount Pinatubo eruption, as did Rabaul vulcanologists in 1994, both areas being evacuated in

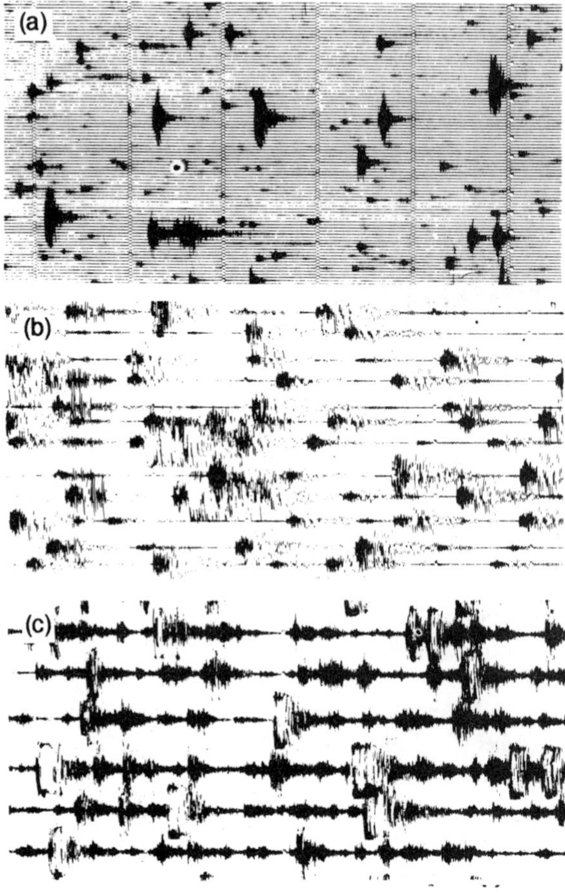

Figure 9.1 Seismograms from near Mt Pinatubo volcano, the Philippines. Minute marks are visible: (a) Numerous short-period events, presumably from rock fracture; (b) longer-period events which may be from magma movements; (c) closely spaced short-period tremors after the great eruption of June 15, 1991. (Courtesy US Geol. Survey)

time. PC-based seismic networks with automatic location and magnitude determination have been developed (US Geological Survey, 1993).

Magma bodies beneath craters are identified as low-velocity regions by seismic refraction, reflection and travel-time-delay methods (Iyer, 1984; Rundle *et al.*, 1985). Seismic (Figure 9.2) and geodetic data show the existence of a complex magma reservoir beneath Kilauea's summit on Hawaii Island and a magma conduit down to 30 km or more (Wright and Pierson, 1992).

9.2 TSUNAMIS (SEISMIC SEA WAVES)

Tsunamis are long-wavelength sea waves caused by large shallow under-sea earthquakes with vertical movement or slumping which accompany the subducting motion under island arcs, or by under-sea volcanic eruptions and pyroclastic flows into the sea; a prime example of the latter being the famous Krakatoa explosion of 1883 in

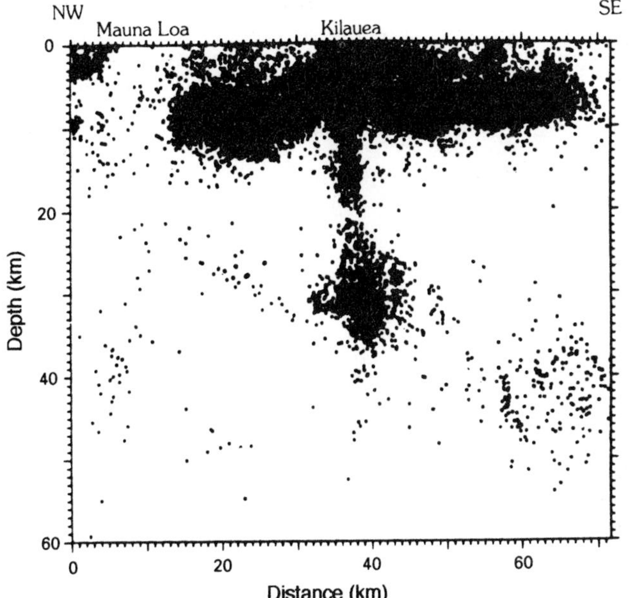

Figure 9.2 NW–SE cross-section of seismicity through the Kilauea volcano, Hawaii. A conduit through which magma ascends appears to be defined. (Courtesy US Geol. Survey)

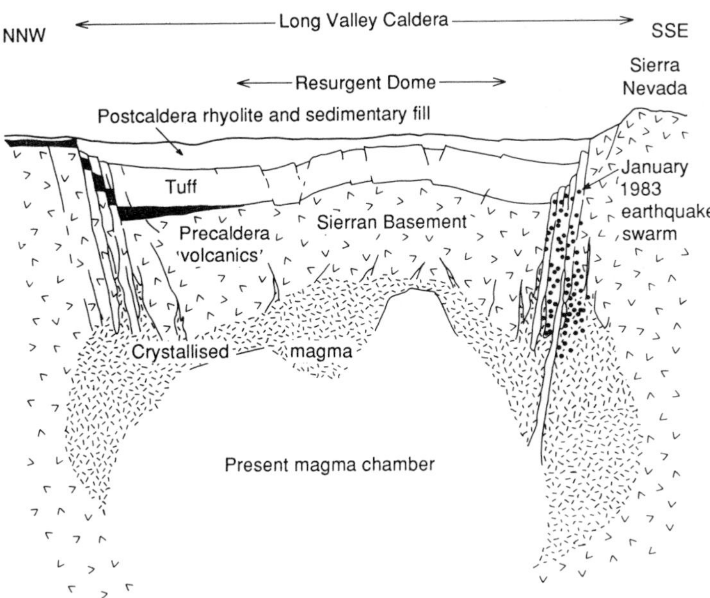

Figure 9.3 A section through the Long Valley Caldera showing the underlying magma chamber inferred from geological and geophysical data, including velocity data. Black dots represent earthquake foci. (After Hill *et al.*, 1985)

Indonesia (Ward, 1989). This produced a tsunami over 30 m high in some shallow coastal areas, causing several thousand deaths, and was recorded as far away as Australia and South Africa. Tsunamis are popularly and incorrectly called tidal waves, but there is no connection with the tides. However, the long wavelengths, which may be hundreds of kilometres, cause slow tidal-like sea movement. Note that pure strike-slip (horizontal) faulting would not result in a tsunami.

These ocean surface waves are gravity waves, i.e. the restoring force is gravity, not elasticity as in ordinary seismic waves. Their velocity depends on the depth of water h, and the acceleration of gravity g, for wavelengths much greater than water depths, as is usual,

$$V = (gh)^{1/2}$$

In the deep oceans they attain speeds up to 790 km/h, but have quite small amplitudes of less than a metre. They are therefore not noticed on board a ship in deep ocean. However, approaching a coastline, the velocity decreases as the depth decreases and the energy of the wave is preserved by increasing its height (apart from energy lost by bottom friction). A 1 m high wave in the open ocean would increase in height to about 3 m in water of depth 50 m and to 5 m in a depth of 10 m. The greatest heights occur in funnelled bays and estuaries. Milne reported one of 210 ft (64 m) in Kamchatka in 1737 (Ward, 1989), and that at the Flores Island, Indonesia, in 1992 reached 26 m with perhaps 2000 deaths. Since tsunamis are commonly produced by a downward movement of a subducting plate, the first movement is often a withdrawal of the sea, an important warning of the crest to come.

Most tsunamis are produced in the Pacific Ocean, bordered as it is by many subducted coasts, but some occur in the Indian and Atlantic oceans, such as in the disastrous Lisbon earthquake of 1755 which affected the coasts of Portugal, Spain and Morocco. The most tsunami active region is the Japan–Taiwan region (Iida and Iwasaki, 1983). A 1983 Japan Sea earthquake tsunami killed 100 people and in 1993 an earthquake and tsunami off south-west Hokkaido in Japan, with a vertical run-up of up to 30 m, caused 197 deaths (Hokkaido Tsunami Survey Group, 1993; Tanioka *et al.*, 1993). Also in 1993 there was a disastrous tsunami on Flores Island, Indonesia (Yeh *et al.*, 1993). In 1994 a tsunami of up to 15 m height killed over 60 in the Philippines. A disastrous tsunami on the Nicaraguan coast in 1992 killed about 170 and destroyed over 1500 homes. This tsunami reached 1 km inland with a maximum run-up of 10 m. It was produced by a slow thrust earthquake large for its magnitude, M_s7.0 (Kanamori and Kikuchi, 1993).

Tsunami damage occurs on steep coastlines, as in Japan from the great tsunami of 1896, 24 m high, which devastated the north-east coast of Honshu, drowning 26 000 people, and also in Japan from the great Chilean earthquake of 1960 on the other side of the Pacific. Tsunami travel great distances because of their low attenuation. On Hawaii Island the same event produced a 9 m tsunami in the town of Hilo (Figure 9.4) and in 1946 an Aleutian earthquake caused 159 Hawaiian deaths. Hawaii has suffered 16 damaging tsunamis since 1819.

Figure 9.5 shows the travel-times of tsunami waves across the Pacific. Fortunately, as the body seismic waves are much faster, there

Figure 9.4 The tsunami of 1946 at the mouth of the river at Hilo, Hawaii. Part of the bridge has already been destroyed by an earlier wave. (After *Pacific Science*, V.1, No. 1, Macdonald *et al.*)

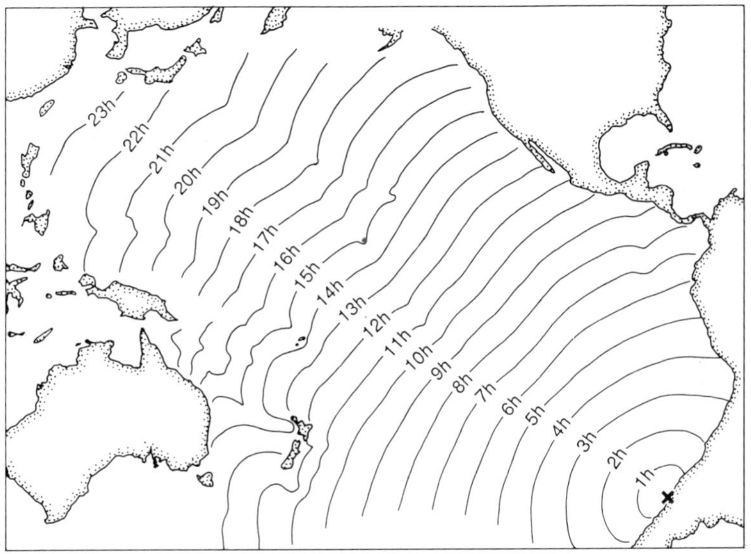

Figure 9.5 Travel-times of tsunamis across the Pacific Ocean from the 1960 Chilean earthquake. Refraction is caused by changes in ocean depth. (After Hisamoto and Murayama, 1961)

is often time to provide warning to populations if the earthquake is not too close. There are international warning systems based on the quick detection of shallow under-sea earthquakes. The Pacific Tsunami Warning Center in Hawaii can issue Pacific-wide bulletins within 15–20 min. The Japanese have an additional system for their region.

Under-sea earthquakes also cause underwater avalanches on continental slopes, known as turbidity currents, which strongly affect sediment distribution. The P waves from submarine earthquakes can be felt on nearby ships as a jolt, as if the ship had hit a rock. Submarine eruptions also produce long-ranging seismic waves which travel through the 'Sofar' low-velocity wave guide in the ocean (Figure 9.6). These high-frequency waves arrive after the normal body waves and are called T (tertiary) waves (Walker, 1984).

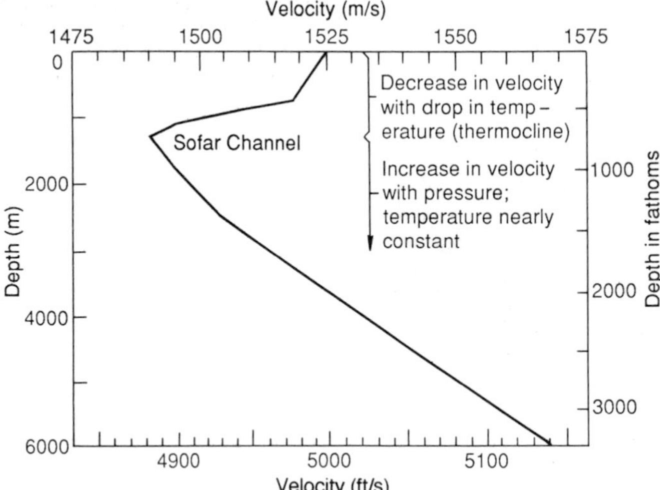

Figure 9.6 The SOFAR low-velocity channel in the deep ocean, showing typical velocity values. (After Sheriff, 1984)

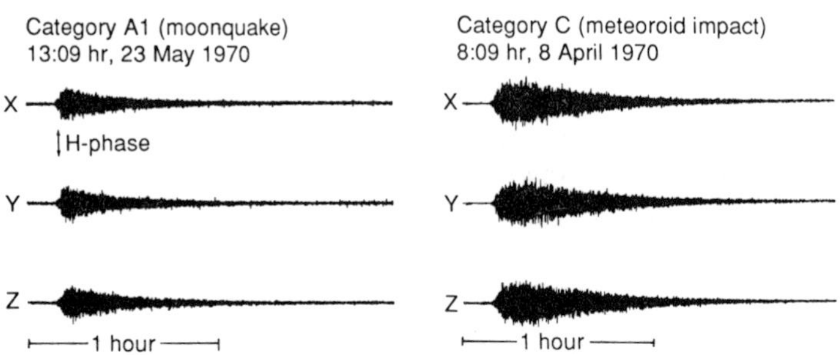

Figure 9.7 Seismograms of a moonquake and the impact of a meteoroid. (After Latham *et al.*, 1971)

Earthquakes can also cause oscillations in lakes and bays called seiches. For a square lake of width L the oscillation period is

$$T = \frac{2L}{(gh)^{1/2}} \text{ seconds}$$

The great Lisbon earthquake produced oscillations in European lakes, including distant Sweden.

9.3 MOONQUAKES

Seismographs were placed at three locations on the Moon during the US Apollo space missions in 1969 and later. A seismic network operated for eight years, with data automatically radioed back to Earth. The Moon has a very low seismic attenuation and the deep regolith of dry, broken material produced by impacts of large meteorites during the Moon's history scatters the seismic waves

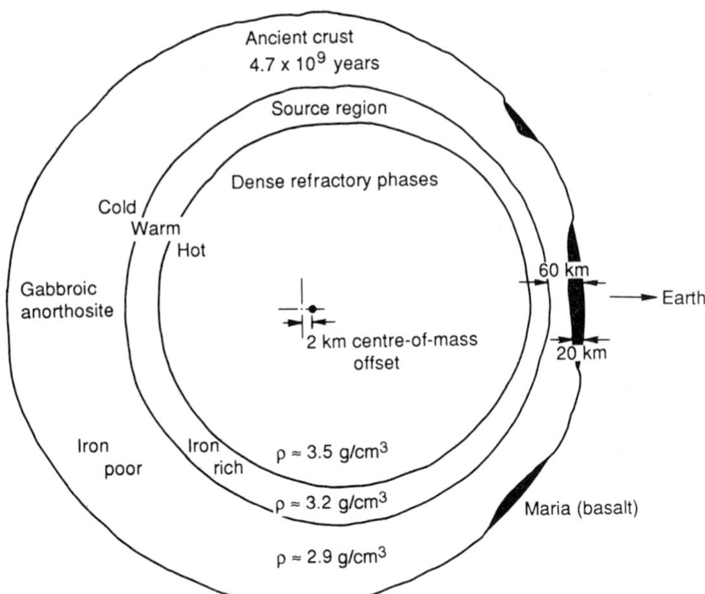

Figure 9.8 Schematic cross-section of the Moon showing possible structure from gravity data. Note offset position of centre of mass. (After Anderson, 1974)

strongly so that the records appear quite unlike normal seismograms (Figure 9.7). With the lack of lunar atmosphere, background noise (microseisms) was extremely low and the instruments were able to operate at very high gain.

They showed a much lower seismicity than here on Earth: about 1000 moonquakes per year, nearly all below magnitude 2, with a total annual energy release of only 10^{15} ergs (Toksoz *et al.*,1977; Lognonne and Mosser, 1993). This is because of the lack of plate tectonics on the Moon which is a colder body and has a very thick lithosphere. These small moonquakes were detected mostly from deep within the Moon (700–1000 km) and correlated with the solid tide on the Moon produced by the gravitational pull of the Earth as the Moon rotates monthly. Other shallower events may have been due to faulting in the crust of the Moon (Lognonne and Mosser, 1993). The seismometers also detected tremors produced by impacts of small meteorites on the lunar surface, about 100 per year. Artificial impacts made by parts of the spacecraft systems (no longer needed) crashing onto the Moon were useful as they landed at known times and places allowing velocity determination. One impact made the Moon reverberate for over two hours. Such data assist in the determination of models of lunar structure (e.g. Anderson, 1974; Figure 9.8).

A seismograph was also placed on planet Mars in 1977 in the Viking landing but was of low sensitivity and recorded no marsquakes; however, the faulting visible in satellite photographs has led to their forecast.

9.4 NUCLEAR EXPLOSIONS

Since underground nuclear tests began in the 1960s, seismology has provided the most important method of detecting and locating them and distinguishing them from earthquakes. Much research has taken place in the USA, the UK and the former USSR towards the detection of underground nuclear explosions which have been limited to a yield of 150 kilotons by the Threshold Test Ban Treaty (Bolt, 1976; Dahlman and Israelson, 1987; Sykes, 1987; Richards and Zavales, 1990; Ringdal, 1990; Stump, 1991; Woods and Helmberger, 1993). The worldwide seismograph network was set up largely for their detection.

The records can be difficult to differentiate from earthquake records unless above about M4.5, but it is believed that explosions above 5–10 kilotons can be reliably monitored seismically (unless decoupled from the ground in a cavity). An explosion of 10 kilotons produces an event of Richter M4.8; there are about 1500 natural earthquakes of this size each year.

Characteristics recorded include wave spectra, amplitude ratios, focal mechanisms, locations and depths. One of the main methods of discrimination is the comparison of magnitudes determined from P waves and surface waves, including for high-frequency waves (up to 30 Hz). Nuclear explosions have been useful in providing large energy sources at accurately known positions and times, the largest almost 10^{18} joules (Figure 4.5), for travel-time calibration and for crust–mantle refraction surveys (e.g. Doyle, 1957; Carder *et al.*, 1966; Scheimer and Borg, 1984).

Chapter 10
Earth structure and the continental crust

10.1 INTRODUCTION

The Earth was once thought to be a largely liquid body with a 'crust' as an outer shell. Descartes originated the idea of a crust plus a metallic centre in the 17th century. A high-density central region for the Earth was suggested in the 19th century from the great contrast between the density of crustal rocks (2.5–2.9 g/cm^3) and the average Earth density of 5.5, known since the famous Cavendish-Michell experiment of 1799 in which the Earth was 'weighed' (i.e. its mass determined) by measuring the gravitational constant G. Also, the Earth's moment of inertia, known from the rate of precession to be only 80% of that for an Earth of uniform density, indicates a sharp increase of density with depth. Hopkins (1839) and Kelvin (Thomson, 1863) studied the tides in the Earth (the solid tides) and showed that the Earth is largely a solid body.

In the 19th century the discovery of two classes of meteorites, the stony and the iron meteorites, then believed to be remnants of a shattered planet (now thought to be from much smaller asteroids), led to the suggestion that the Earth was made up of a stony silicate mantle over an iron core. Wiechert (1896) extended Hopkins's and Kelvin's discussions of mean Earth rigidity and suggested that the core was non-rigid to explain the rather low mean rigidity μ of the Earth (the liquid core produces an amplification of the Earth tide response). This mean rigidity is about that of steel, which is actually small for such a large globe of gravitationally compressed material. Jeffreys (1926a) clinched the argument by using seismic, tidal and Chandler wobble data. It is also believed that there must be a fluid core to make possible the generation of the Earth's magnetic field by magnetohydrodynamic processes. Brush (1980) discusses the history of the discovery of the core.

Oldham (1906) detected a discontinuity at the core boundary using early earthquake records, and in 1909 Mohorovičić first recognized

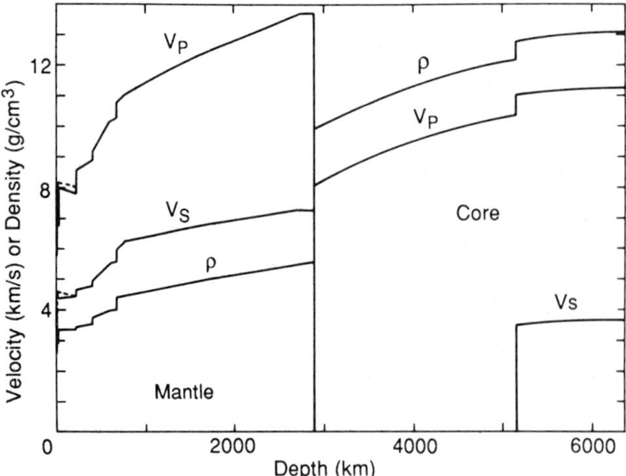

Figure 10.1 The Preliminary Reference Earth Model (PREM) of Dziewonski and Anderson (1981). Dashed lines show the higher velocities of horizontal components in the upper mantle (anisotropy). (After Dziewonski and Anderson, 1981)

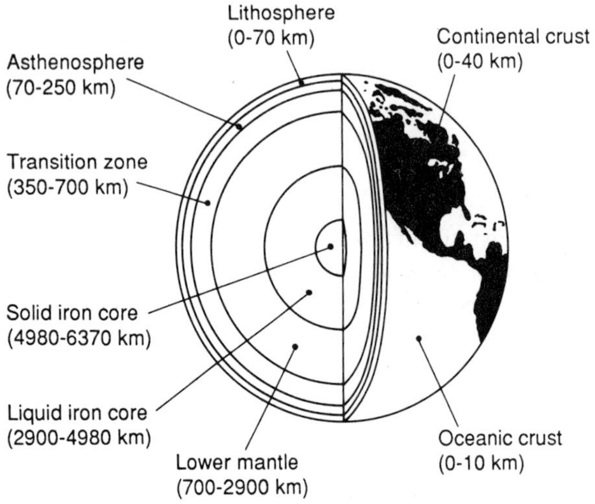

Figure 10.2 Earth structure with a solid iron inner core, liquid iron outer core, mantle, asthenosphere, and lithosphere which includes the crust. (After Press and Siever, 1986)

crustal phases on seismic records. Seismological research in the 20th century has determined average velocities and densities through the Earth (e.g. Figure 10.1) with a remarkable accuracy of better than 1% (Lay *et al.*, 1990). The travel-times of P and S waves were converted to values of velocity versus depth for a spherically symmetrical Earth. This was made possible by the mathematical method of Herglotz and Wiechert (see Bullen and Bolt, 1985). The data prove clearly that the Earth has a crust, a mantle and a core (Figure 10.2) which geochemical and density arguments show are chemically different (Bullen, 1975; Anderson, 1989; Dziewonski, 1989).

We can liken the Earth to a blast furnace with the lighter slag at the top, the crust, and the iron at the bottom, the core (Anderson, 1989). Heat in the Earth was produced by radioactivity and the release of gravitational energy during the accretion of the Earth (its origin by accumulation of material) and by the separation of the core from the mantle. This has resulted in convection at rates of centimetres per year and upwellings a few hundreds of degrees hotter than the subducting downwelling slabs which cut the upper mantle. Seismology provides information on thermal differences as cold material has higher velocity than warm material. Three-dimensional images of the interior now span from the crust to the Earth's inner core (Dziewonski and Woodhouse, 1987; see next section). This recent development of seismology has thrown light on various surface features and also on the directions of mantle convective flow (through velocity anisotropy, p. 22). Convection through the whole mantle seems likely, but would be inhibited by the phase changes in the upper mantle and so possibly in both layers (p. 117).

10.2 SEISMIC TOMOGRAPHY

This is an important development in seismic imaging used both in the study of the Earth's interior and in applied seismic exploration in which the travel-times along many ray paths from different directions crossing the region of interest are used to produce three-dimensional models by mathematical inversion (Aki *et al.*, 1977; Nolet, 1987; Romanowicz, 1991; Phillips and Fehler, 1991; Iyer and Hirahara, 1993) (Figure 10.3). It is also called 3D inversion. Both body waves and surface waves may be used on scales from local to global. To study a particular area a set of distant sources is used to give differential travel-times (corrected for distance and local heterogeneity) over a network of stations covering the area of interest. An example is the recording of shots from one borehole in another hole (cross-borehole) to give velocities and structure (Lines *et al.*, 1991).

For global studies a worldwide group of earthquakes and stations are needed, although resolution is rather poor for this because of the locations of stations and epicentres. It has become more practicable in recent years with the increased number of digital seismograph stations around the globe and greater accuracy of data (Anderson and

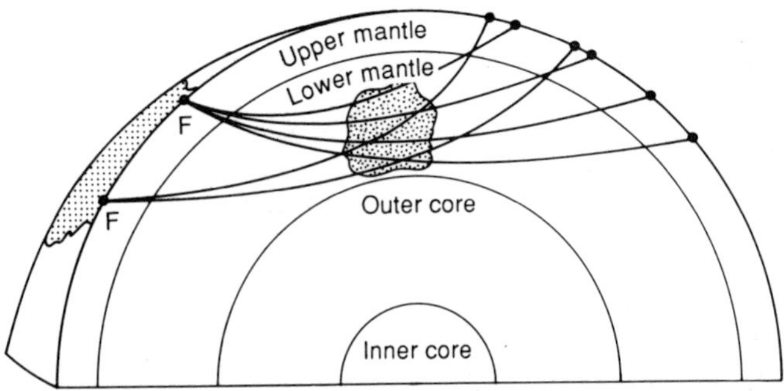

Figure 10.3 The principle of seismic tomography which can also be on a local scale. Many ray paths are required through each region of interest. (After Bolt, 1993)

Dziewonski, 1984; Thurber and Aki, 1987; Montagne and Tanimoto, 1991; Lay, 1994). Digital seismographs are necessary with a wide frequency response and high dynamic range. This tomography has enabled nominal resolution of the spherical harmonics of Earth structure to degree 36.

Data are entered into a computer and treated in a similar way to data used in medical tomography to image the body. The region (or whole Earth) is broken up into a large number of blocks of different tectonic type and given initial velocities. With many rays passing through each block mathematical methods, such as least squares, may be used to adjust the initial velocities so as to match the measured travel-times. The three-dimensional images of the Earth obtained have resolutions limited by the number of paths passing through each block into which the Earth is divided. Details of the interior (as velocity anomalies) can then be seen which may be related to such surface features as mid-ocean ridges, volcanic areas and 'hot spots' such as Hawaii and Iceland, and subduction zones (Figure 1.9). Seismic velocities also give constraints on estimates of density and viscosity and the shape of the geoid.

Another important method now used is the mathematical inversion of seismic *waveforms* to produce models of Earth structure.

10.3 THE CONTINENTAL CRUST

The 'crust' has come to be defined by seismic velocities. It is the outer part of the Earth above the Mohorovičić discontinuity ('Moho' or M) which is the level at which the P velocity increases to 7.8–8.2 km/s (Figures 1.7 and 10.4), and is found almost everywhere (Meissner, 1986). Below it is the mantle, and to a first approximation the crust, or more exactly the lithospheric plates, behave as if floating on the

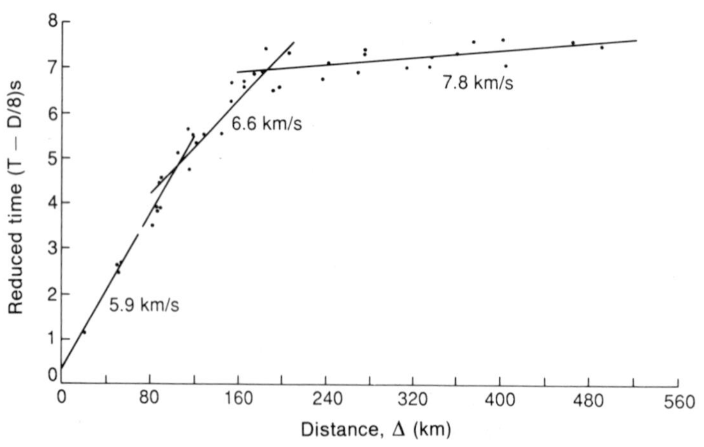

Figure 10.4 Travel-time curves for a crustal survey. Note the time is plotted as a reduced travel-time, t-distance/8 km/s, to give a clearer set of curves. (After Finlayson, 1968)

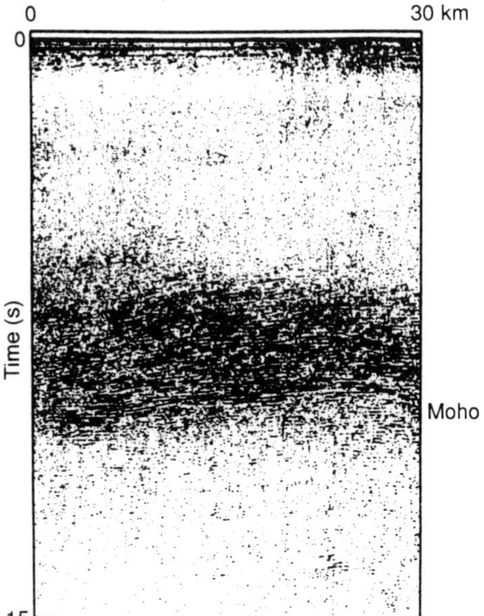

Figure 10.5 Combined (stacked) reflection data over continental crust showing lower-crustal layering and the Moho at the base of the layering. (From Warner, 1990)

mantle. The M discontinuity, the second most prominent boundary in the Earth's interior, is itself less than a kilometre thick in some areas but appears to be a few kilometres thick in many others. However, this definition of the base of the crust, based on velocities, is imprecise in some localities and is open to some debate. Although many refraction surveys around the world have verified the existence of this increase in velocity (Figure 1.7), more recent reflection recordings do not suggest a simple structure. In some cases a single reflector at the Moho is indicated; in others there appears to be a double reflector or a number of reflectors with no certain discontinuity level (Figure 10.5). Thus there is still room for questions about the nature of the M discontinuity (Griffin and O'Reilly, 1987; Oliver, 1988; Jarchow and Thompson, 1989).

The crust is not to be confused with the lithosphere, which is usually defined by its long-term rigidity and not by velocity (see p. 113 and Figure 10.2). The lithosphere is usually thought of as the outer comparatively brittle portion of the Earth that makes up the 'plates' of plate tectonics and includes the crust and part of the upper mantle.

The continental crust comprises about 30% of the Earth's surface, the rest being oceanic, but the crust as a whole is only 1% by volume of the Earth and less than 0.4% by mass. It is penetrated by drilling to only a fraction of its depth, but makes up almost all the material studied by geologists. Hence the importance of geophysical methods, particularly seismology, in the study of the Earth.

There are three major crustal types: continental, oceanic and transitional crust. The transitions from continent to ocean are usually difficult to define as they are under the seas or under thick sedimentary

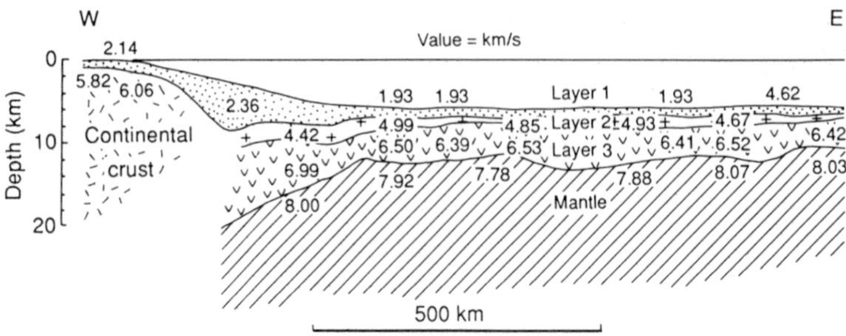

Figure 10.6 Ocean crustal structure and the ocean–continent transition from refraction surveys in the Atlantic, east of Argentina. (After Ewing, 1965)

basins (Figure 10.6). Crustal structure and thickness measurements are made by both seismic refraction and reflection surveys, the latter more recently being used on a wide scale, often with common mid-point (CMP) or wide-angle methods (p. 147) (Braile, 1991). Refraction data provide values of broad velocity at different depths and regions while reflection data provide more detailed structural information. Sources have been explosives, including nuclear (e.g. Scheiner and Borg, 1984), well-located earthquakes, Vibroseis on land and air-guns at sea (pp. 149 and 153). Studies of S-wave velocities also have increased as the ratio V_p/V_s and Poisson's ratio can give more information on lithology.

Programmes of seismic refraction studies of the crust and upper mantle have been carried out on all continents over the last 40 or more years (e.g. see Leven *et al.*, 1990; Pakiser and Mooney, 1989), partly as a development from earthquake studies. Reflection surveys of the deep crust were developed more slowly, but surveys in many countries have now produced about 30 000 km or more of land and ocean reflection data (Barazangi and Brown, 1986a,b; Pakiser and Mooney, 1989; Finlayson *et al.*, 1990; Mooney and Meissner, 1992). Klemperer and Hobbs (1992) describe the basics of acquisition and the processing used and show results of the BIRPS surveys.

Crustal thicknesses on the continents usually range from 30 to 45 km, with greater values where the Moho dips down under some mountain ranges such as the Alps and the Andes, to as much as 60 or 70 km deep. The Himalayas and Tibet region appears to be underlain by about double the normal crustal thickness, produced by the underthrusting of the Indian continent. Continental crusts are believed to have grown by the accretion of island arcs, e.g. the accretion of New Guinea onto the Australian continental mass (Meissner and Mooney, 1991). Oceanic areas are about 60% by area of the crust and 20% by volume, the crustal thicknesses are significantly thinner, normally ranging from 5 to 15 km, increasing from the young mid-ocean ridges towards the deep oceans (Figure 1.8). However, oceanic plateaux and aseismic ridges can have thicknesses exceeding 30 km.

Reflection profiling is highly effective in the marine environment because of the excellent coupling of the air-gun sources to the water

(Mooney, 1989). But vertical reflections from deep continental depths cannot always be obtained; vertical or near-vertical reflections can only be obtained if the M discontinuity is sharp relative to wavelength. According to Mooney and Meissner (1992), the Moho is commonly 3–5 km wide and is possibly made up of thin layers. Reflection amplitudes are small, perhaps 10 000 times weaker than the surface waves, but larger beyond the critical distance because of total reflection (Figure 2.15). On land, large explosive sources are more effective than vibrators (Brocher and Hart, 1991).

S-wave reflections, which are now being attempted, are somewhat different to P-wave reflections, showing no 'transparent' upper crust, perhaps because of fluid-filled cracks (Fuchs *et al.*, 1990). For interpretation of data, as well as the comparison with that calculated from models (the 'forward method') refraction and reflection data may be 'inverted' by sophisticated mathematical techniques to determine structure (Huang *et al.*, 1986; Pavlis, 1989). Inversion has advantages (Zelt and Smith, 1992).

As well as the use of body waves in refraction and reflection, surface waves from earthquakes are useful in crustal and upper mantle studies, particularly in remote areas such as Antarctica. Inversion of surface-wave dispersion is used to give S-wave velocity structure. Surface-wave dispersion curves (Figure 2.7) can also be fitted to curves derived from crustal/mantle models using the known P and S velocities from refraction surveys to give structure. Phase velocities across station networks (Figure 10.7), or group velocities to individual stations are used (see p. 21).

Seismic P velocities in the upper crust beneath the sediments range from about 5 km/s (and less at shallow depths) above about 8 km depth where pores and cracks affect velocity, to between about 5.8 and 6.3 km/s at deeper levels where the greater pressure closes such voids. Refraction surveys in some areas have been interpreted to indicate a discontinuity (the Conrad discontinuity) at 15–25 km

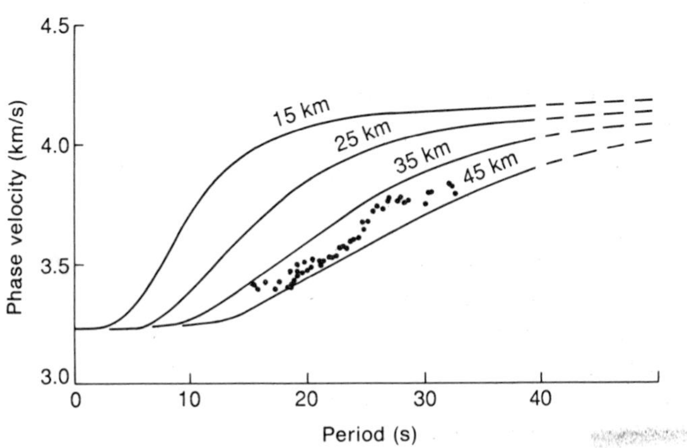

Figure 10.7 The use of the phase velocities of surface (Rayleigh) waves to determine crustal thicknesses. (After Ewing and Press, 1950)

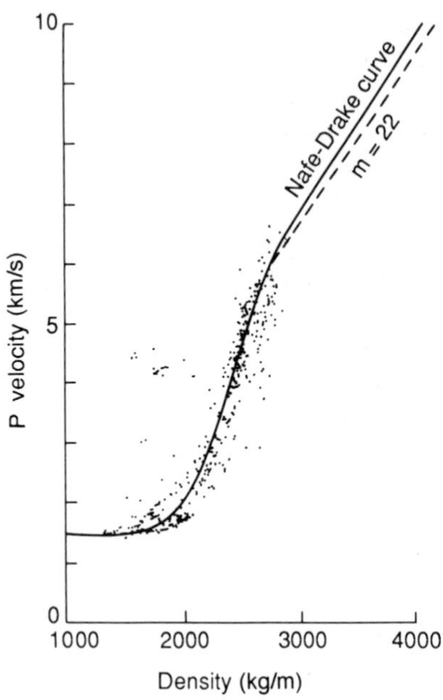

Figure 10.8 The Nafe–Drake curve of P velocity versus density for saturated sediments, sedimentary rocks and hard rocks. Birch's empirical curves for mean atomic weights m are also shown. (After Nafe and Drake, 1963)

depthwhere the P velocities increase to 6.5–7 km/s. Such levels have been taken as the bottom of the upper crust. However, other surveys have not found such a discontinuity and more recent reflection surveys and deep drilling (see p. 104) show much more complication and heterogeneity than a simple two-layer structure for the crust (Litak and Brown, 1989).

There are three main causes of velocity variations in the Earth: (a) pressure and temperature gradients with depth, (b) changes in the chemical composition of the rocks, and (c) phase changes affecting mineralogy but not composition (e.g. at the 400 and 650–700 km levels in the upper mantle). Assuming an average crustal density of 2.9, the lithostatic pressure increases with depth at a rate of 30 megapascals (MPa) per km, giving a pressure of 103 MPa or so at the bottom of the crust. Together with the estimated temperature increase with depth in the crust, this enables estimation of P and S velocities in the crust from laboratory measurements for comparison with field data. A useful correlation between P velocity and density is shown in the Nafe–Drake curve (Figure 10.8) for saturated sediments and igneous rocks. Velocity anisotropy (p. 22) can be valuable in providing information on rock fabric (Fountain and Christensen, 1989).

10.4 UPPER CONTINENTAL CRUST COMPOSITION

The relation between seismic velocities and crustal rock type has been studied in laboratory measurements (e.g. Fountain and Christensen,

1989). However, such correlations are inexact and depend on such things as metamorphic grade. The ratio of P velocity to S velocity and Poisson's Ratio are more reliable indicators. The continental crust composition differs from the oceanic crust in being rich in granitic and associated rocks (quartz and feldspar being the common minerals), whereas the oceanic crust is largely made up of basic rocks, basalts, dolerites and gabbros. The top continental 'layer' is a sedimentary one of variable thickness: from zero in parts of the shields and averaging about 0.5 km over shields, up to 10 km in deep geosynclines and basins. In young fold belts the average is about 5 km of sediments. Beneath the sediments is the strongly folded and metamorphosed basement or upper crust, locally intruded, especially by granites and granodiorites (Figure 1.7). Young compressional belts, such as the Alps, have deep crustal roots.

It was at first thought that the continental crust was made up of a granitic or 'sialic' upper layer plus a basaltic lower crust. This fitted the early seismic velocity models. However, geochemical estimates of average upper crustal composition suggest material intermediate between granites and basic rocks, granodiorite or tonalite (Anderson, 1989). Also, basement rocks of the upper crust have densities averaging 0.1–0.15 higher than granite, as we see from negative gravity anomalies over true granite batholiths. Thus granite cannot make up the average crustal rock. Deep erosion, as much as 15 km, suggested for some ancient shields from metamorphic studies, apparently exposes samples of deeper crustal lithology (Salisbury and Fountain, 1990).

Deep seismic reflection surveys in several countries, such as COCORP profiling in the USA, BIRPS in the UK and the 'Lithoprobe' transects in Canada, have recorded reflectors at various levels in the crust (Brown *et al.*, 1987; Mathews and the BIRPS Group, 1990; Meissner and Brown, 1991). COCORP profiling in the south-eastern USA has shown dipping reflectors that may mark a former suture between North America and Africa. However, continuous reflections over distances longer than a few kilometres are rare, and amplitudes and dips may vary considerably within sections. Some areas have distinctive reflectors in the upper crust, produced possibly by intruded sills, metamorphic layering, alternating felsic and mafic layers, or fluids (e.g. Hyndman, 1988; Meissner and Wever, 1992), and some can be related to surface fault zones. But in other regions, particularly with granite batholiths, few reflectors have been detected in the upper crystalline crust, being comparatively transparent. In the same region the lower crust may have a number of reflectors terminating at about Moho depths (Figure 10.5), particularly in extensional terranes (see below).

Cratonic areas often show deep sedimentary basins in the upper crust and dipping reflectors at upper and lower depths. In several areas major sutures and terrane boundaries have been imaged as dipping reflectors. For example, in central Australia a thrust forming the boundary between exposed granulites and amphibolite facies has been imaged to a depth of at least 30 km (Goleby *et al.*, 1989).

Russian workers have drilled the deepest hole in the crust so far using a hydro-powered drill, and reached a depth of over 14 km in the Kola Peninsula (Kozlovsky, 1987). They found sediments much deeper than expected and the drill was still in the 'upper' crustal material at 13 km. Boundaries, such as the Conrad discontinuity, interpreted in seismic surveys, were found to actually correlate with metamorphic grades. Mineralized fissures were encountered much deeper than previously thought and there was a large influx of fluids at various depths. The deep presence of sediments may explain why average velocities are not higher than the usual 5.9–6.3 km/s range. This work has taken place over 20 years. Another super-deep hole being drilled in Germany (Fuchs *et al.*, 1990) has reached 9.1 km, and another is planned in the USA.

10.5 THE MIDDLE AND LOWER CONTINENTAL CRUST

Refraction surveys result in P velocities for the middle continental crust of 6.0–7.1 km/s, commonly 6.5–6.8 km/s (Figure 1.7) (Holbrook *et al.*, 1992). For the lower crust, P velocities are between 6.4 and 7.5 km/s (Percival and Berry, 1987; Mereu *et al.*, 1989; Holbrook *et al.*, 1992). The lower crustal velocities can be divided into two groups, 6.7–6.8 and 7.0–7.2 km/s, and the effects of magmatism or high-grade metamorphism have been suggested to explain these high velocities. Observed velocities have also been ascribed to dry granulitic rocks in the lower crust (granodiorite–diorite) or possibly amphibolite, if in a wet state. Holbrook *et al.* (1992) argue for a bulk mafic lower crust and point out that lower crustal reflections are common in rifts and extended crust where mafic compositions are also likely from the high Poisson's ratios.

Samples of possible former lower crust are seen in deeply eroded Precambrian terrains, in uplifted portions, and in xenoliths of volcanic rocks (Dawson *et al.*, 1986); also in granulite terrains, such as in the Ivrea Zone of the European Alps, the Adirondack Highlands of New York State, or the Frazer Range of Western Australia. Many granulites have a suitable velocity range and density (about 2.8) and low radiogenic heat production, but so do some mafic rocks.

From deep reflection studies, Brown *et al.* (1987) suggest the possibility of sediments in the lower crust. Data show heterogeneous lower crusts with major low-angle thrusts and apparent low-velocity lamellar structures several hundred metres in length (Figure 10.5). The discontinuous subhorizontal reflections recorded from depths of 20–40 km are variously interpreted as mafic lenses, sheets, fluids, metasediments or shear zones (Newton, 1990; Braile, 1991). Underplating and intrusion of the lower crust by mantle-derived mafics has been suggested by Warner (1990), so that some lower crusts could be younger than the upper crust. High velocities in the lower crust have also been noted in volcanic terranes. The Mohorovičić discontinuity in some areas seems to be a planar structure between the lower crustal reflectors and the relatively transparent upper mantle.

Chapter 11
Oceanic areas and margins

11.1 CONTINENTAL MARGINS

The continental margins are important in understanding the history of the continents; they are also where many geological resources are found. The margins are the sites where the continents may grow by plate interactions incorporating new sediment and/or volcanics. Thus parts of the continents consist of the remnants of ancient margins. Continental margins may be divided into 'Atlantic' inactive or divergent type coasts formed by continental rifting, such as around the Atlantic and Indian Ocean coasts, and 'active' subducting (or convergent) margins (Pacific type), as along many converging Pacific coasts, e.g. the west coast of South America and the south coast of Alaska (Talwani, 1989). Island arcs also mark a subducting oceanic plate but have a marginal ocean basin beyond.

The divergent type margins have low seismic activity and a change in crustal thickness from continental to thinner oceanic, in some cases starting abruptly at about the 200 m ocean depth contour, but in many cases the position of the true ocean–continent contact is unknown (Bott, 1982). Much of the change in crustal thickness may occur under the continental slope, the comparatively steep edge of the continental shelf (Figure 10.6). Most of the inactive (divergent) margins are covered by thick sedimentary basins and thus are targets for hydrocarbon exploration. The sediments are divisible into those formed during continental rifting and those deposited during drift when sea-floor spreading is under way. Seismic refraction and reflection surveys indicate that some inactive continental margins have thick high-velocity (over 7 km/s) lower-crustal material extending beneath both the sedimentary troughs and adjacent ocean basins (e.g. Trehu *et al.*, 1989), supporting the idea that initiation of sea-floor spreading is accompanied by massive volcanic activity, intrusion and underplating by large volumes of melt.

Active subducting margins are characterized by zones of seismicity (Wadati–Benioff zones), sometimes double zones, dipping beneath island arcs (Figures 5.1 and 5.2) or Andean-type continental edges.

These zones mark the descent of oceanic lithosphere in the plate tectonic subduction process. The plates are largely affected by 'slab-pull' downwards as they cool to higher densities. Some of the upper sedimentary layers, particularly poorly consolidated waterlogged material, may be scraped off the descending plate by the overriding plate. These sediments appear to pile up to form an accretionary prism while the lower consolidated sediments are subducted into the mantle. However, in other areas all the sediments may be accreted, or all subducted (Mutter, 1986).

11.2 OCEANIC CRUST

In oceanic areas the mobile plates or lithosphere are from 50 to 150 km thick and include the crust and part of the upper mantle. One of the most important discoveries of geophysics, made in the 1950s, was the contrasting crustal structure under the oceans compared to the continents (Figures 1.8 and 10.6). This breakthrough was made largely by Ewing and his co-workers at Lamont Geological Observatory (NY) and Hill of Cambridge University (UK) with refraction surveys at sea. The oceanic crust is much thinner, about 7 km on average (plus the ocean), and much younger as shown by the magnetic anomaly dates (another important breakthrough). Under the sediments (if any), the ocean floor and crust is made up of basaltic-type rocks (Table 11.1) produced at the mid-ocean ridges where separating plates decompress the asthenosphere below, allowing it to rise and melt. The greater part of the oceans was formed in the last 200 million years compared to over 3000 million years for the oldest continental areas, the oldest ocean crust so far being 170 million years old Jurassic sediments from the western Pacific found in the International Ocean Drilling Program.

However, the first indications that oceanic crusts may be thinner were from the surface-wave dispersion studies of Gutenberg (1924) and the series of pendulum gravity observations made by Vening Meinesz in submarines in the 1920s and 1930s. Meinesz showed that the oceans are in approximate isostatic equilibrium with the continents, except for subducting zones (trenches and island arcs) where large anomalies exist. The general isostatic balance is explained by an Airy-type model, i.e. the thinner ocean crust is balanced by the shallower higher density (3.3 g/cm^3) mantle beneath. In fact, a simple Airy model predicts an oceanic crustal thickness of 6.6 km for an ocean basin 5 km deep, compared to measured thicknesses of 6–7 km by refraction work, i.e. it balances a 35 km continental crust for densities of 2.90 g/cm^3, 3.30 g/cm^3 and 1.03 g/cm^3 for crust, mantle and ocean, respectively.

Multichannel seismic arrays deploying thousands of hydrophones in a streamer perhaps 2–3 km long (Figure 11.1; see Chapter 16 and Mutter, 1986) have been used in all the oceans to map the sediments

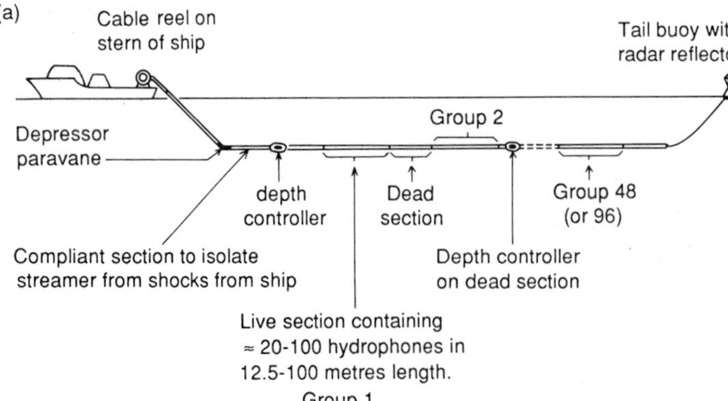

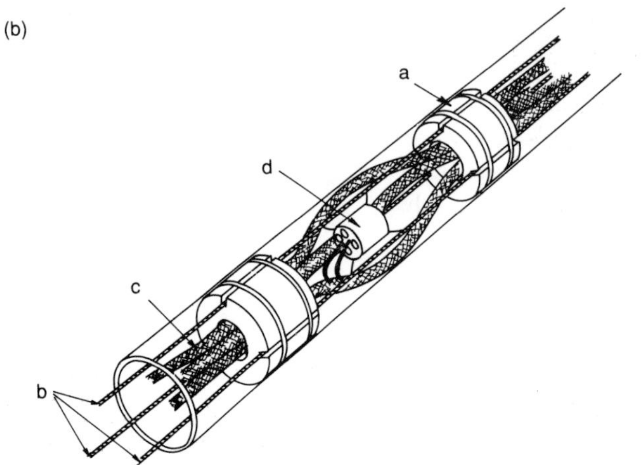

Figure 11.1 The seismic streamer used in marine reflection surveys. (a) A seismic streamer, typically 3–5 km long, used in marine reflection studies; (b) detail of the streamer showing a) plastic spacer, b) cables, c) bundle of conductors, and d) a hydrophone. (After Seismic Engineering Co.)

and the igneous crust. As for gravity and magnetic work, ocean seismic surveys are much quicker than on land, a ship steaming day and night at 5 knots can cover distances of hundreds of kilometres in a few days. Kennett (1977) discusses the mathematical methods used in the interpretation of such ocean surveys.

Initially explosives were used for the energy sources but groups of air-guns (emitting sound pulses of frequency 6–60 Hz) are the common source today (Figure 11.2). For example, in the widely separated surveys carried out in the Pacific by Talwani *et al.* (1982) and others, sharp reflections were obtained from the M discontinuity (e.g. Figure 11.3). Surveys have also shown other refractors and reflectors in the sediments and igneous parts of the oceanic crust, which is in three 'layers': layer 1 consists of sediments; layer 2 consists of basaltic rocks as pillow lavas (layer 2A) and sheeted dykes (layer 2B); and layer 3 is the main 'oceanic' layer, apparently largely gabbroic (see Table 11.1 and Figure 10.6).

Figure 11.2 A seismic survey vessel towing air-gun sources. (Courtesy of Sean Waddingham and Horizon Exploration Ltd)

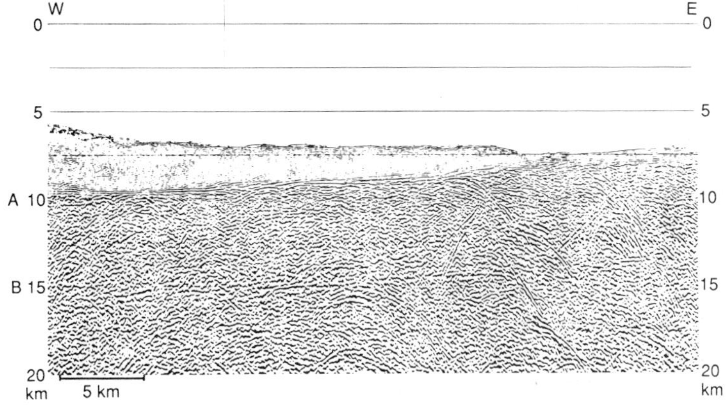

Figure 11.3 A reflection section across the Japan Trench showing the subducting crust. A is the top of the subducting crust and B the Moho. (After Matsuzawa *et al.*, 1980)

Arrivals from layer 1 are only rarely first arrivals due to the small velocity contrast with the ocean above and the thinness of the layer. Layer 1 sediments are often absent on the crusts and flanks of mid-ocean ridges because of their youth and the lack of time to form sediments, but are perhaps a few kilometres thick in ocean basins (Franchetau, 1983). Reflection surveys have added detail on the sediments which may be further divided into an upper section of unconsolidated sediments down to a prominent reflector 'A', and a

Table 11.1 Characteristics of oceanic crust

	Composition	Velocity (km/s)	Average thickness (km)
	Water	1.5	4.5
Layer 1	Sediments	1.6–2.5	0.4 (0 on ridges to 5 km in basins)
Layer 2	Pillow basalts	4–6	1.5 (0.7–4)
Layer 3	Gabbroic?	6.4–7	5 (4–7)
Upper mantle		7.5–8.3	

lower one of consolidated sediments. Deep-sea drilling has revealed the poorly consolidated clays and turbidites above this reflector and highly lithified formations such as chalk and limestone below it (Mutter, 1986). The most prominent reflector below the seabed (which itself is a very strong reflector) is the layer 1–layer 2 boundary where sediments overlie basaltic rocks. This boundary is consistently rougher than the seabed above, which is smoothed out by the sedimentation (Figure 2.17).

Layer 2 is a thin but important zone of pillow basalts from volcanic eruptions on the sea-floor and sheeted dykes intruded beneath. Layer 2 outcrops at the mid-ocean ridges where there is no sediment and has been sampled, drilled, and photographed by deep diving expeditions. It is important as a major source of the ocean magnetic anomalies. P velocities vary from about 3.6 to 6 km/s, the low velocities occurring in the ridges and volcanic centres where hydrothermal circulation of seawater occurs.

Layer 3 is the main oceanic layer with an average thickness of about 5 km and P velocities of 6.4–7.7 km/s. It is thought to consist of gabbroic rocks, the remains of magma chambers, which also provide some of the magnetization of the oceanic crust. Layers 2 and 3 are formed by injection of magma as the lithospheric plates move apart.

A unique partnership of 18 nations has pooled finance and expertise to explore the structure and history of the ocean floor using the research drill ship *Joides Resolution*, managed by the Texas A & M University. The ship can drill to depths of 2 km into the ocean crust in up to 8 km of water. This International Ocean Drilling Program (IODP) and its predecessor the Deep Sea Drilling Project (DSDP, 1968–1983), have obtained cores beneath marine sediments, the deepest yet being Hole 504B on the Costa Rica Rift (Becker *et al.*, 1989; ODP Leg 148 Shipboard Scientific Party, 1993). This hole penetrated 274 m of sediments and (recently) 1800 m of sub-basement to 2111 m below the sea-floor, to near the bottom of the oceanic crust. Recovered core indicated about 575 m of extrusive lavas over about 200 m of transition into over 500 m of intrusive sheeted dykes. In such work the ages of the oceanic crust predicted by the magnetic

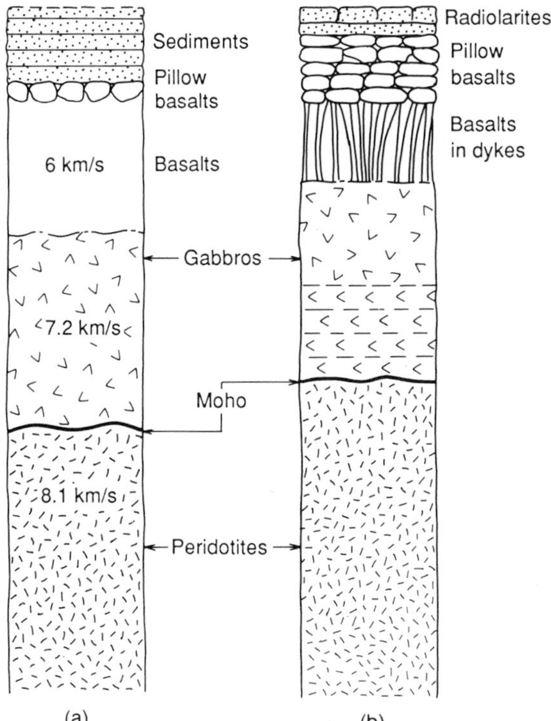

Figure 11.4 (a) Oceanic crustal structure from seismic data. (b) Oceanic crust as determined from field observations of ophiolites. Velocities are also used in correlating ophiolites with oceanic structure. (After Allegre, 1988)

anomalies have been confirmed and the important time-scale of magnetic reversals.

The oceanic upper mantle has a P velocity averaging about 8.1 km/s, as is commonly found under the continents, but this increases with distance from the ridges and thus with the age of the ocean crust. This Pn velocity appears to be anisotropic; for example, being 8.3 km/s perpendicular to the ocean ridges, and so along plate drift directions, and 8.0 parallel to the ridges (Figure 2.9) (Christenson and Salisbury, 1975). The presence of oriented olivine in peridotitic material could explain such anisotropy.

The thrusting of ophiolites up onto land in a few locations such as Cyprus also probably shows us sections of oceanic crusts (Figure 11.4), although they may represent crust formed near island arcs rather than at mid-ocean ridges (Anderson, 1989; Parson *et al.*, 1992).

11.3 THE OCEAN RIDGE SYSTEM

The ocean ridge system is one of the largest structures on the globe (Figures 1.8 and 5.3), rising 2–3 km above ocean basins and 500–1000 km wide, with a total length of 74 000 km. The ridges are believed to have been formed by rifting of the entire lithosphere, not just the crust. After separating, the plates then eventually drift apart and hot

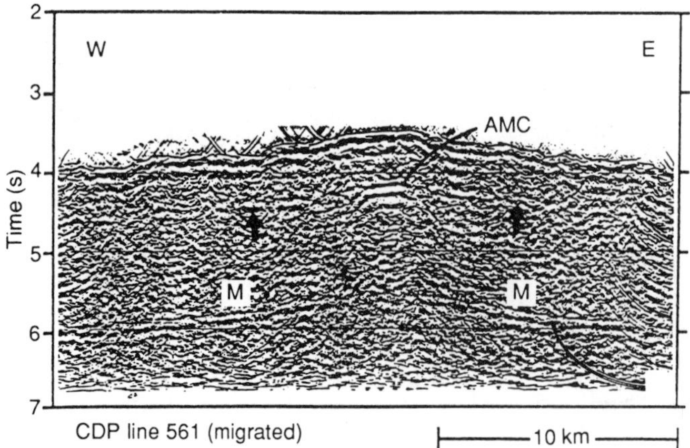

Figure 11.5 A reflection profile across the East Pacific Rise. The strong reflector beneath the rise axis is believed to be the top of the axial magma chamber (AMC). Reflections from the Moho (M) can be seen to within a few kilometres of the axis. The arrows indicate a reflector interpreted as the top of a frozen magma chamber. (After Sinton and Detrick, 1992)

material rises from the mantle to fill the gap. The new plate or lithosphere so formed then slowly cools from about 1000 °C and so contracts by as much as a few kilometres as it moves away from the mid-ocean ridge, forming the deep ocean basins (Cox and Hart, 1986). The ridge topography is very rugged, with median rift valleys along the crests of the slower spreading ridges (<3.5 km/year). These rift valleys are typically more than a kilometre deep and as much as 40 km wide (Mutter and Karson, 1992). Transverse fracture zones are also prominent, cutting across the ridge systems and displacing them with transform faulting. The presence of the ridges worldwide was first indicated by earthquake epicentre maps. Later, earthquake mechanism studies showed normal faulting along the ridges and strike-slip movement along the fracture zones, thus indicating the directions of plate motion (Isacks *et al.*, 1968).

Refraction measurements reveal low Pn velocities and other studies low S velocities beneath some ridges. Surface waves show high attenuation (low Q), presumably because of magma chambers and high temperatures. Reflection surveys have been similarly interpreted (Figure 11.5). Kent *et al.* (1990) found from detailed reflection surveys evidence that the mid-ocean ridges beneath the East Pacific Rise are fed by chambers with widths of only 800–1200 km, i.e. less than previously believed (Sinton and Detrick, 1992). Data suggest the shallow (~ 1.5 km) magma chambers exist only beneath the axes of fast spreading ridges, probably because of the different thermal conditions (Morgan and Chen, 1993). Massive sulphide deposits produced by volcanic action have been located on the Mid-Atlantic Ridge and other ridges.

Chapter 12
Lithosphere and mantle

12.1 THE LITHOSPHERE AND ASTHENOSPHERE

The lithosphere (from the Greek *lithos*, meaning 'rock') is the outer, more brittle part of the Earth, originally called the crust, now defined by the seismic velocities. The lithosphere includes the crust and part of the upper mantle (Figure 10.2). It has been likened to an eggshell cracked into sections (the tectonic plates). There is uncertainty in its exact definition as rigidity depends on the temperature, stress and time-scale involved (Maxwell, 1984; Anderson, 1995; Drummond, 1991). More strictly, the lithosphere has a range of brittleness, the upper 20 km or so being the seismogenic region where most earthquakes occur, apart from the subduction zones (Fuchs and Froidevaux, 1987). The existence of the lithosphere has also been explained in rather simple models of viscous convection in the mantle of high viscosity (about 10^{22} poises, like glass at room temperature) and high Rayleigh numbers ($\sim 10^7$), with the lithosphere being the cold boundary layer.

Under the young ocean ridges the lithosphere is only about 5 km thick or less (excluding the ocean). Under the ocean basins, the lithosphere is probably 50–150 km thick. ScS data show a reflector about 60 km under the western Pacific (Revenaugh and Jordan, 1991). There is some evidence that the lithosphere is much deeper under very old continental terrains, possibly more than 300 km (e.g. Lerner-Lam and Jordan, 1987; Jordan *et al.*, 1989).

In many areas there is evidence of the asthenosphere (from the Greek *asthenos*, meaning 'weak') from body-wave times and surface-wave dispersion. The asthenosphere is beneath the lithosphere where velocities are lower and the temperature probably approaches melting. Gutenberg (1959) first gave body-wave data, and Lehmann (1964) later presented evidence for a velocity discontinuity at a depth of 200 km that may be at the bottom of a zone of partial melting (Hales, 1991). The lower velocities were confirmed by other workers measuring the dispersion of surface waves (Dorman *et al.*, 1960) and the inversion of 400 modes of oscillations of the Earth (Anderson and

Hart, 1976). S velocities, in particular, appear to be lower at about the same depths.

The asthenosphere is important as its higher temperature and ductility may allow the slow movement of the lithospheric plates above, and the production of basaltic melts, at least in some areas. The depth to the asthenosphere (i.e. the thickness of the lithosphere) is difficult to measure or estimate, as both temperature and lithology in the upper mantle are only inferred. Its depth is often taken as 60–200 km. Below that the upper mantle is only as ductile as ordinary glass, but still ductile enough to flow over geological time (White and McKenzie, 1989).

12.2 THE MANTLE

The mantle makes up the greatest part of the globe, 83% by volume and 69% by mass, extending from the base of the crust (the M discontinuity) to the core and including the asthenosphere and part of the lithosphere (Figure 10.2) (Lay, 1987; Lay *et al.*, 1990). The mantle has been studied using body waves, surface waves, oscillations of the Earth and recently by seismic tomography (Dziewonski and Woodhouse, 1987; Lay, 1994).

After great earthquakes, trains of long-period surface waves circle the globe many times. These waves provide worldwide averages of upper mantle structure through measurement of their dispersion (group and phase velocity versus period; Figure 2.7). Since the late 1960s, free oscillation data have provided more direct information on densities in the mantle, although no great changes were made to radial Earth models (Bullen and Bolt, 1985). More accurate travel-times to large distances became available using nuclear explosions (e.g. Carder *et al.*, 1966) and accurate measurements of regional times using conventional explosives (e.g. Green and Hales, 1968).

The upper mantle is defined as from the base of the crust down to a depth near either 410 or 660 km (usually the latter) where velocity 'discontinuities' are located (Figure 12.1). At 410 km a comparatively steep velocity increase from about 8.5 to 9.3 km/s occurs. This was first noted by Jeffreys (1970) and then confirmed by Niazi and Anderson (1965), who also showed the presence of a discontinuity near 670 km, using one of the first seismograph arrays to trace rays in the mantle (Figure 12.2). Whereas the crust–mantle (M) discontinuity is accepted as a compositional change, the 410 km discontinuity is commonly believed to be due to a phase change of olivine to a higher-density modified spinel structure (Ringwood, 1975). The upper 400 km of the mantle has the largest lateral velocity variations, up to 10% for S velocities, and these variations are associated with surface tectonic provinces (see p. 117).

The region between the 410 and 660 km levels is often denoted the

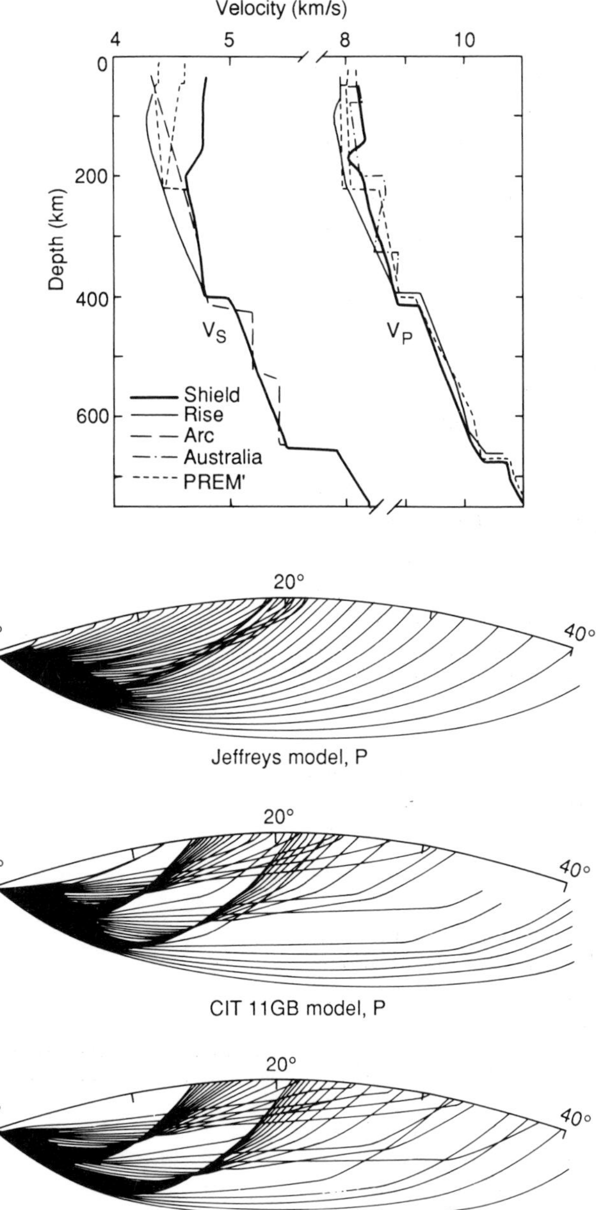

Figure 12.1 (a) S and P velocity profiles for the upper mantle in different tectonic regions. Shields generally have faster velocities than other regions between about 100 and 200 km depths. (After Press and Siever, 1986)

Figure 12.2 Ray paths through the upper mantle (based on older velocity data) showing the ray concentrations at 14° and 21° caused by the velocity increases near 400 and 670 km. (After Julian and Anderson, 1968)

'transition' region, and the discontinuity at 660 km depth (which is a velocity increase of again about 5%) as the base of the upper mantle. It is uncertain whether this lower discontinuity is caused by a phase change to a higher density mineral (including perovskite) or a chemical boundary separating the upper and lower mantles, although the former explanation seems to be preferred (Anderson, 1989;

Menzies, 1990; Revenaugh and Jordan, 1991). This level is also about the lower level of earthquake occurrence, possibly because of the temperature and pressure reached by the subducted oceanic plates at that depth causing loss of brittleness, but other explanations have been proposed. The 410 and 660 km discontinuities appear to have topographic variations of up to 40 km, as seen in reflection data.

Shearer (1990) has used the Global Digital Seismograph Network to stack long-period data and show that the 410 km and 660 km discontinuities are worldwide, plus a weaker transition at 520 km, possibly caused by a change from a modified spinel structure to spinel. The 520 km discontinuity and one at about 220 km depth had been noted by others, including Revenaugh and Jordan (1991) who used the ScS waves strongly reflected from the core. The 220 km step may be the bottom of the asthenosphere or zone of partial melting but appears not to be worldwide.

The high-quality digital global networks that began operation in the 1970s, together with the development of computers, have made possible three-dimensional modelling of the Earth using great amounts of data (Aki *et al.*, 1977). Dziewonski and Anderson (1981) obtained a Preliminary Reference Earth Model (PREM) using 1000 oscillation periods, 500 travel-times and 100 seismic attenuation values, the latter because they affect free oscillation periods. PREM is believed to be the definitive average Earth model, a good model of the elasticity of the Earth.

Dziewonski and Woodhouse (1987) and Montagne and Tanimoto (1991) have published global contours of velocities at various levels in the mantle and core using seismic tomography, combining the travel-time data for many body-wave paths through the Earth and surface-wave and oscillation data.

As Jeffreys pointed out, the Earth's mantle must flow as it cannot sustain masses such as mountain ranges higher than about 10 km. The mantle is undergoing thermal convection driven by heat from the

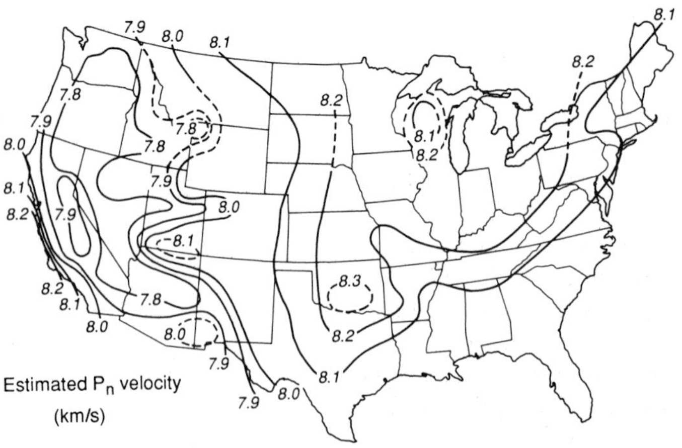

Figure 12.3 Estimated P_n velocities beneath the USA. (After Tucker *et al.*, 1968)

core and from internal radiogenic heat. There is much discussion on the amount of interaction there is between the upper and lower mantle convections, i.e. whether there is whole mantle convection or separate systems in the two sections. Creager and Jordan (1986) claim they have evidence of some penetration by subducted plates into the lower mantle (i.e. below 660 km) and Olson *et al.* (1990) propose that whole mantle convection is consistent with seismic tomography, mantle geochemistry and numerical modelling of convection. There does not seem to be a positive answer to this question yet, although the 660 km boundary does seem to be difficult to penetrate. See also Peltier (1989) and Jordan *et al.* (1989).

12.3 REGIONAL VARIATIONS IN THE MANTLE

The onion-type layered models of the Earth with radial symmetry are believed to be 99% correct, but there are important differences in P and S velocities from region to region, first clearly shown by refraction studies using nuclear and other explosions in the 1950s and 1960s (e.g. Herrin and Taggart, 1962). Iyer and Hitchcock (1989) give a more recent account. Figure 12.3 shows the regional variations in P (P_n) velocities in the topmost mantle beneath the USA. Similar variations occur in Europe, Australia and elsewhere. Cleary and Hales (1967) demonstrated the different delay times for P times at large distances to different regional structures. Doyle and Hales (1967) found larger delays for S waves, suggesting higher temperatures in the upper mantle for recently active areas (see also Souriau and Woodhouse, 1985).

Surface waves are sensitive to shear-wave velocities in the upper mantle. Velocities are fast under shields but slow under young, active regions and oceans, particularly mid-ocean ridges, as confirmed by tomographic methods (p. 97). This involves breaking the Earth up into many blocks, classifying them according to their tectonic history, e.g. ocean basins of different ages, Precambrian cratons, Phanerozoic sediment platforms and active tectonic provinces.

These methods reveal a large-scale pattern of low velocities (hence likely high temperatures) correlating with the locations of hot spots and thus convection in the upper mantle compatible with surface plate motions (Olsen *et al.*, 1990; Romanowicz, 1991). Velocities are low near extensional features such as oceanic ridges, continental rifts and back-arc basins, but high in the old cooler shields and ocean basins. Relatively thick, high-velocity and cool roots appear to exist under stable cratons, perhaps as deep as 400 km (Lerner-Lam and Jordan, 1987). Azimuthal anisotropy of velocities in the mantle beneath oceanic areas has been interpreted as being due to flow alignment of crystals such as olivine (Ribe, 1989).

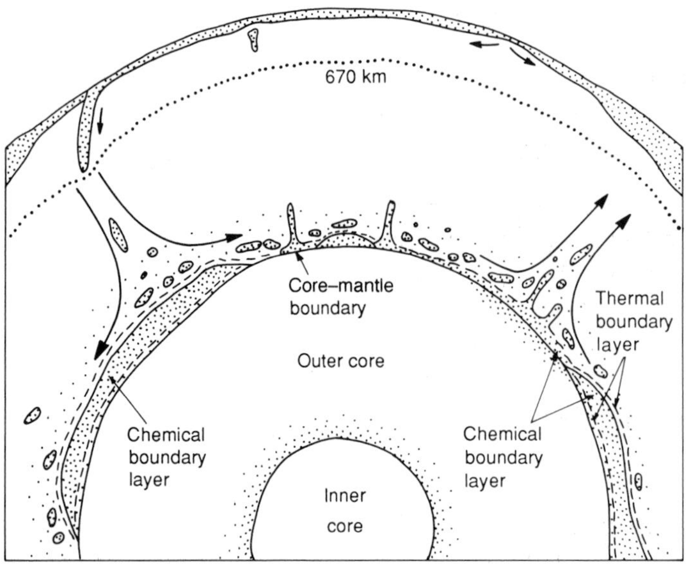

Figure 12.4 The core and the core–mantle transition zone and possible large-scale mantle convection. (After Lay, 1989)

12.4 THE LOWER MANTLE

The lower mantle may be defined as extending from the base of the transition zone (about 660 km) to the core–mantle boundary at a depth of about 2885 km. As for the upper mantle, seismic arrays have proved valuable in measuring travel-time gradients ($\mathrm{d}T/\mathrm{d}X$) directly, and therefore more accurately (e.g. Wright and Cleary, 1972). Velocities increase smoothly with depth down to within 300 km or less of the core where the velocity gradients are reduced to zero and lateral velocity variations are comparable to the upper mantle.

This lowest portion of the mantle was earlier examined by Bullen, who named it the D″ region. It is believed to be an inhomogeneous transition zone of greatly varying thickness up to about 300 km; it is not well understood, but it is probably a major thermal boundary layer and laterally heterogeneous (Figure 12.4). One explanation is that it is cold slab material that has sunk to this depth. Another is that it may be a chemical reaction zone between the liquid iron core and the solid silicate mantle (Knittle and Jeanloz, 1991) and is perhaps the most active zone of the Earth (Jeanloz and Lay, 1993). The temperature at the base of the lower mantle has been estimated as 2550–2750 K. The core–mantle boundary seems smooth and sharp, but appears to have significant (10 km) long-wavelength topography (Young and Lay, 1987; Loper and Lay, 1995).

Seismic tomography has also shown that there is a correlation between long-wavelength shear velocities in the lower mantle and the geoid, thus indicating that convection is occurring there (Dziewonski and Woodhouse, 1987).

12.5 CONSTITUTION OF THE MANTLE

There has been much discussion about the chemical constitution of the mantle (Ringwood, 1975; Anderson, 1989). Seismic velocities and discontinuities have been compared with high-pressure laboratory experiments on elastic properties of minerals.

Liu (1985) has concluded that the mantle (and therefore about half the Earth) is made up of perovskite-type ferromagnesian silicates. This conclusion is based on high-pressure laboratory experiments in which this material has been synthesized, and by comparison of its density and elastic properties with the values for the mantle from seismology. Xenoliths which have been brought up in kimberlite pipes indicate that the uppermost mantle down to depths of 150 km has peridotite as the common rock type. This is composed mainly of the mineral olivine which is believed to make up at least 65% of the upper mantle. Creep in olivine is thought to be the main deformation mechanism and is important in convection models.

Chapter 13
The core

The Earth's core, with a radius close to 3485 km (at a depth of 2885 km), is larger than Mars (3397 km) and is 55% of the Earth's radius and 31% of its mass (Figure 12.4). Its temperature has been estimated as over 4000 K, with a recent estimate of 6000 K. Thus the outer core is a white-hot iron 'ocean', with a viscosity perhaps close to that of water. The inner core is solid with a density of 12.5–13 g/cm^3 (Jacobs, 1987; Masters and Shearer, 1990).

Convection in the iron liquid outer core appears to be much faster (a few kilometres per year) than in the mantle and produces the Earth's magnetic field. The solid inner core has a radius near 1221 km, somewhat smaller than the Moon (1738 km) and is probably growing as crystals settle out of the outer core.

The core–mantle boundary is the most distinct boundary in the Earth, with contrast in properties across it greater than those between air and rock at the surface. It was discovered not long after the first reliable seismographs were built, early in the 20th century. Both it and the inner core boundary are quite sharp. Oldham (1906) first noticed that P waves from antipodal earthquakes were late compared to the extrapolated time curves (Figure 1.5). He proposed a core of lower velocity, which might produce a shadow zone, to explain this.

In 1913 Gutenberg verified this predicted shadow zone (see Gutenberg, 1959), one for P between distances of 105° and 145°, with a sharp drop in P amplitudes beyond 100° (Figure 1.5). He also accurately measured the depth of the core–mantle boundary. This depth (2885 km) can also be determined from PcP and ScS reflections (Figure 1.2), as originally used by Jeffreys to confirm Gutenberg's determination. Earth oscillation data also suggest a very low-rigidity outer core. As well, the large amplitudes of observed SKS waves (S waves converting to P (K) through the core) necessitate a liquid outer core, as a solid core would greatly reduce the conversion of energy at the core boundary. SKS suffers a small change only in velocity in converting from S to K and so does not refract much (Figure 1.2).

The early low magnification seismographs did not detect P waves beyond about 105°, until at 143° where short-period P waves appear strongly again, but delayed (Figure 1.5). The waves are semi-focussed by refraction through the mantle and outer core to give maximum amplitudes, a 'caustic', at 143°. Rays are also refracted to between 143° and 188°, depending upon the angle of the ray from the focus. The 143° ray is the deepest through the outer core; the shallowest goes to 188° distance. The S waves do not reappear at 143°. These facts are explained by a core of P velocity 40% lower at the mantle–core boundary and with no S. Note that no such PKP and SKS rays have paths through the topmost core region. Velocities there have to be indirectly inferred.

In the 1920s improved seismographs and earthquake location made it possible to detect some small P arrivals in the 'shadow zone' between 110° and 143°. Diffraction was at first thought to explain these observations, but Jeffreys showed that diffraction could only occur for a few degrees from 143° distance. A Danish seismologist, Inge Lehmann (1936), suggested the now-accepted explanation that PKIKP waves sharply refract at a solid inner core boundary (Figure 1.2) (I is a P wave in the inner core). These waves transmit to between 180° and 110°.

Shear velocities and densities within the core are determined mainly from Earth oscillation data and the strongest evidence of significant rigidity of the inner core is from such data, particularly those with significant amplitudes in the inner core (Bolt, 1987). Core models suggest an inner core S velocity of 3.5 km/s. In 1972, S waves in the inner core (PKJKP) were also claimed to have been detected; however, this is uncertain. Other evidence came from the huge LASA seismic array of 525 centrally recording seismographs in Montana, USA, which in 1970 recorded PKiKP waves reflected from the inner core boundary from Nevada underground nuclear explosions. The amplitudes of the reflected PKiKP waves suggested an inner core density of 12.5 g/cm^3, and recently possible detection of oscillations of the inner core suggest central densities of 13 g/cm^3.

Models of the structure of the core suggest significant complexity (e.g. Morelli *et al.*, 1986; Dziewonski and Woodhouse, 1987; Jeanloz, 1990; Creager, 1992). There is evidence that the inner core is anisotropic, with the axis of symmetry aligned with the Earth's axis. P-wave velocities have been reported by Creager (1992) as being 3.5% faster along axis than for equatorial paths.

13.1 COMPOSITION OF THE CORE

Suggestions have been made in the past that the core is composed of (a) condensed hydrogen, and (b) high-pressure silicates stripped of outer electrons to produce a high-density material of low melting point and high conductivity. Neither idea has significant support today.

Seismic data show that the fluid outer core is 10% less dense than if it were pure iron at that pressure and temperature. The presence of lighter elements such as O, S, Si, K or H alloyed to the iron has been suggested, oxygen being the most popular. High-pressure and high-temperature experiments carried out on likely core constituents using either shock waves or diamond anvil methods with lasers for heating (e.g. Brown and McQueen, 1983; Jeanloz, 1988) have shown consistency with an iron-alloy model (90–95% iron, plus some nickel). Diamond anvil methods can now produce pressures greater than at the Earth's centre. Shock-wave experiments using hypervelocity projectiles to hit samples have provided momentary pressures of over 500 GPa and temperatures of 1000–7000 K. In those experiments both shock-wave and particle velocities are determined. (In shock waves, particle velocities are comparable to wave velocities.) These values give the variation of density versus pressure (the Hugoniot curve). Inner core densities of up to 13.4 g/cm^3 are suggested at the Earth's centre. The densities suggested are slightly less than for iron, so requiring a lighter element to be present as well, and oxygen may make up several per cent by weight according to Ringwood (1983).

Some recent temperature estimates for the centre of the Earth have increased to about 7000 K, almost twice previous shock-wave estimates. The outer and inner core radii depend upon the relations (cross-over) between the melting-point and temperature curves with depth. At the outer core boundary the melting point drops considerably in going from mantle silicates to the Fe–Ni core. The inner core solidification is due to the great pressure, and the fact that the melting point increases faster with pressure than does the adiabatic temperature.

Chapter 14
Seismic exploration and the refraction method

14.1 INTRODUCTION

Seismic exploration methods (particularly reflection) are mainly used in the search for oil and gas, for which they are essential. It would be unusual to make a find today without seismic surveys. They are also used to some extent in mining for coal (Ziolkowski, 1979), but only to a small extent in the search for metals or for mapping igneous structures (Hawkins and Whiteley, 1981; Hajnal, 1983; Dahle *et al.*, 1985). However, refraction surveys, and reflection to a lesser extent, are useful for mapping shallow stratigraphy in engineering, groundwater and environmental investigations (Romig, 1986; Dooley, 1990; Steeples and Miller, 1990; Ward, 1990).

The reflection process is simple in principle, but in practice it is a much more complex and costly process than the refraction method. To map the numerous reflecting beds, one must determine the changes in velocity, the depths and true dips of the beds, and also reduce the interference from surface waves, refractions and noise, which arrive before and during the reflections. Reflected waves never appear as first arrivals, but their advantage is that they do occur at all distances and it is not necessary to record out to great distances for deep penetration, as in refraction. The much greater detail obtainable with reflection surveys has made them indispensable in hydrocarbon surveys and the method is sometimes preferable in shallow mapping of geology and soil (e.g. Hunter *et al.*, 1984; Pullan and Hunter, 1990; Ali and Hill, 1991) and for detecting sinkholes and voids (Miller and Steeples, 1991). However, reflections from shallow depths are more difficult to resolve and require higher frequencies.

Refraction surveys are an important method for investigating the sub-surface and were used in both crustal studies (see p. 135) and oil exploration before reflection surveys, because of their greater practical simplicity. They provide reliable velocity information and are used for weathering corrections for reflection data (p. 161). The refraction

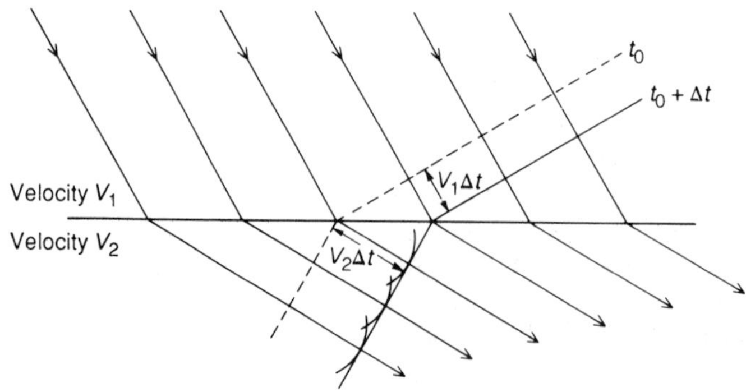

Figure 14.1 Huygen's construction for refraction of a wave at a sharp interface. Each point on the interface may be regarded as a source of wavelets whose envelope is the refracted wave-front

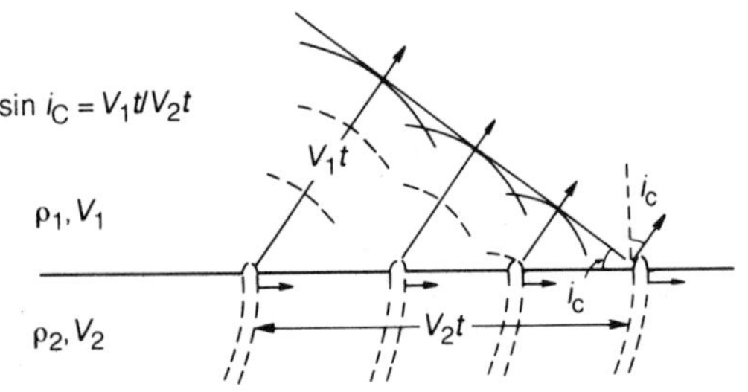

Figure 14.2 The generation of the critically refracted (head) wave by the passage of the wave in the top of the lower, faster layer. (After Griffiths and King, 1965)

method is popular in engineering site evaluation, for example to measure the thickness of soil and softer rock at dam sites and large-scale building constructions (Sjogren, 1984; Dooley, 1990; Langston, 1990; Walker *et al.*, 1991). In hydrology the refraction method is used to map aquifers and sometimes the water-table. However, the principles behind critical refraction are not as simple as in reflection.

14.2 PRINCIPLES OF THE REFRACTION METHOD

Waves are refracted from the top of any layer that has a higher velocity than the layer above (Figures 14.1 and 14.2), and where velocities increase with depth a series of refractions may be produced and returned to the surface (Figure 1.12). Recorded refractions give a simple measure of the refractor velocities if there are clear velocity contrasts (see examples in Musgrave, 1967).

The instruments used in refraction are simpler as often only the first P wave arrivals are measured and many instruments are still analogue. The frequencies of refracted waves are somewhat lower than for reflections, partly because of the greater distances from shot to geophones, thus lower frequency geophones may be used.

Refraction, or bending of the rays, occurs at a boundary between layers of different velocity, as for any wave motion (e.g. light). A wave-front may be constructed at its next position by drawing a series of arcs centred at points on the wave-front, and then drawing the envelope (Figures 14.1 and 14.2). At a boundary an incident wave produces four waves — two reflections and two refractions — except at normal (perpendicular) incidence. A P wave produces a reflected P at a reflection angle equal to the incident angle and an S wave at a smaller angle (Figure 2.15). The converted S wave does not reflect at an equal angle as it has a different velocity to the incident P. There are also two refracted waves, P and S, again with the S wave being refracted at a different angle to the P wave.

Reflection and refraction are governed by Snell's Law,

$$\frac{V_i}{\sin i}\text{(incident wave)} = \frac{V_r}{\sin r}\text{(refracted)} = \frac{V_R}{\sin R}\text{(reflected)}$$

which can be demonstrated by considering the wave-fronts in Figure 14.1. Refracted and reflected waves are produced at the point of incidence, and this point, common to all four waves, moves along the interface at a velocity $V/\sin i$ which defines the component of each wave velocity along that interface, i.e. those components of the incident, refracted and reflected velocities must be equal, as in the above equation which holds for both P and S waves. Note that $\sin i/V$ is called the ray parameter p.

For the case of an incident angle i increased such that the refracted angle $r = 90°$, the refracted wave travels along the top of the *lower* layer (Figure 1.12). The incident angle at which this happens is called the critical angle i_c (Figures 14.3 and 14.4) and

$$\frac{V_i}{\sin i_c} = \frac{V_r}{\sin 90°}$$

or

$$\sin i_c = \frac{V_i}{V_r} = \frac{V_1}{V_2}$$

where V_1 and V_2 are the velocities in the top and bottom media. For

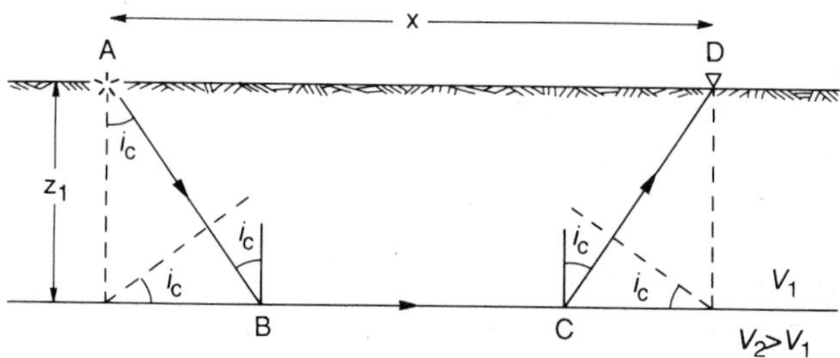

Figure 14.3 Refraction at a horizontal boundary between an upper layer of velocity V_1 and a lower layer of faster velocity V_2

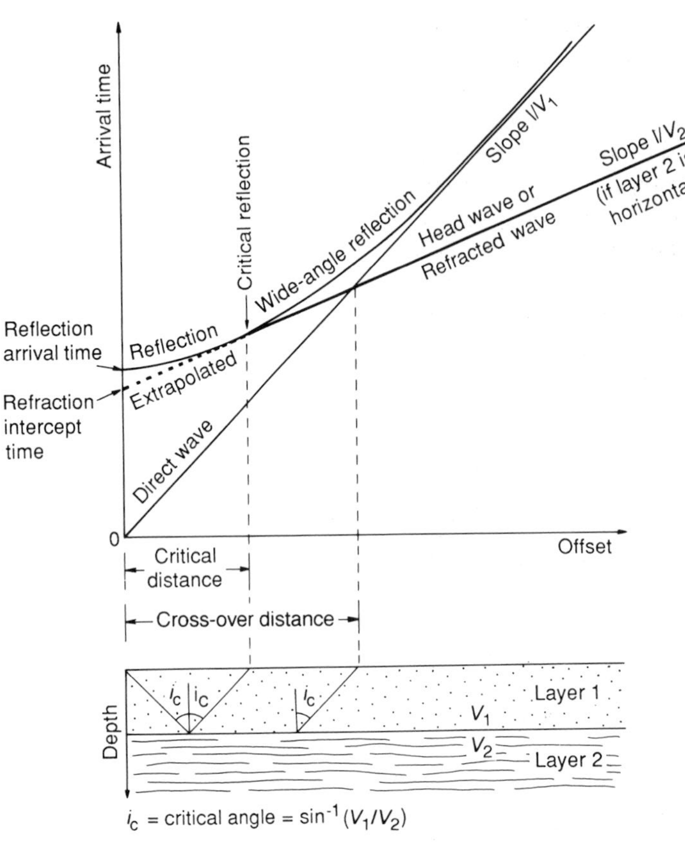

Figure 14.4 The relation between the refraction, reflection, the critical distance (where the critical refraction first appears at the surface) and the cross-over distance. (After Sheriff, 1984)

incident angles greater than the critical angle no energy enters the lower layer and is totally reflected back into the top layer. Thus strong reflections are obtained at distances just beyond the critical distance (Figure 2.15).

The refracted wave disturbs the interface and sends energy back into the upper layer and thus to the surface (Figure 14.2); this is a diffraction phenomenon. (The critically refracted or 'head' wave has an amplitude in the lower layer that decays exponentially with depth below the interface, so it is similar to a surface wave in this respect. The rate of decay depends on the wavelength.) The refraction arrival usually has an emergent character, not as sharp as the direct wave through the top layer. It returns to the surface at the critical angle to the vertical, i.e. the path of least time from source to receiver (Fermat's Principle). An array or 'spread' of seismometers or geophones along the surface will then enable measurement of a velocity from a plot of travel-time t versus distance x (Figures 14.3–14.5). You will note that the extra path-length to further geophones is in the bottom layer and so the slope of the refraction travel-time curve measures V_2.

It is remarkable that apparently a single refracting ray (the critical ray) along the interface radiates energy upwards, when a single ray has theoretically no energy. A theoretical explanation (related to the fact that the incident wave-front is spherical, not plane) was given by Jeffreys (1926b), years after the refraction method had been in use (see also Cagniard, 1939; Cerveny and Ravindra, 1971).

This critically refracted wave (sometimes called the head, conical or lateral wave) is analogous to a bow wave produced by a boat on the water (Figure 14.2), to the 'shock' wave from a supersonic aircraft, and also to Cerenkov radiation. In all these cases the energy source is travelling faster than the natural wave speed in the medium. In our case the refracted wave in the top of the bottom layer is faster than the wave speed in the media above and so sends a bow wave into the top layer.

As can be seen from Figure 14.4, direct waves through the top layer arrive first at shorter distances and the direct travel-time curve is a line through the origin. Refracted waves will arrive at the surface at and beyond the *critical* distance ($X_c = 2z \tan \theta_c$), the minimum refraction distance. The *cross-over* distance is that beyond which the refracted wave arrives before the direct wave which travels through the top layer (Figure 14.4). In refraction work the geophones (seismometers) are usually strung out well beyond the cross-over distance to make the refraction's first arrivals simpler to recognize. Figure 14.5 shows an example of a shallow refraction record with the usual time versus distance plot.

Thus plotting travel-times versus distance will give two curves, the slopes of which define the velocities for the direct and refracted waves in the simple case where the two velocities are quite constant. This simple measurement of velocities at depth is the great advantage of the refraction method. It can be carried out by a single observer, but measurements must also be taken in the reverse direction to allow for dip (p. 131).

Notice also the travel-time curve (a hyperbola) for the reflected wave in Figure 14.4. It arrives after the refracted wave from the same layer but coincides with it at the critical distance. That is why amplitudes are large near this distance. The reflected wave is also asymptotic to the direct wave at large distances, i.e. it eventually merges with it.

The travel-time for a refracted wave on a horizontal interface is (see Box 14.1):

$$t = \frac{X}{V_2} + \frac{2z(V_2{}^2 - V_1{}^2)^{1/2}}{V_1 V_2} = \frac{X}{V_2} + t_0$$

where t_0 is the intercept on the time axis (the delay time) and

$$t_0 = 2z\frac{(V_2{}^2 - V_1{}^2)}{V_1 V_2}$$

Thus the depth to the interface, z is as follows:

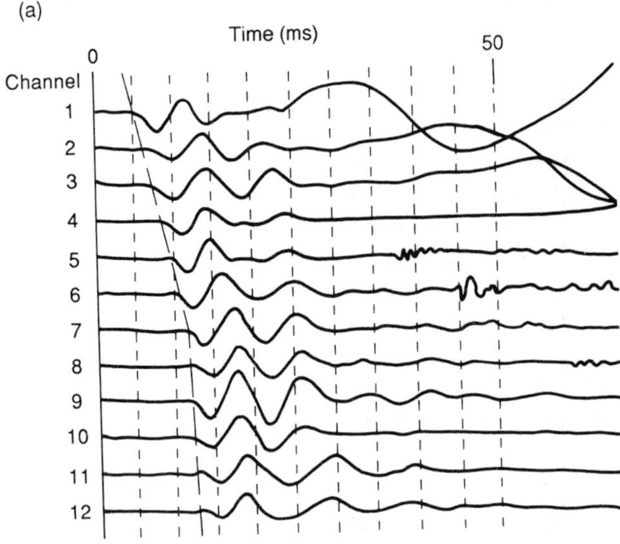

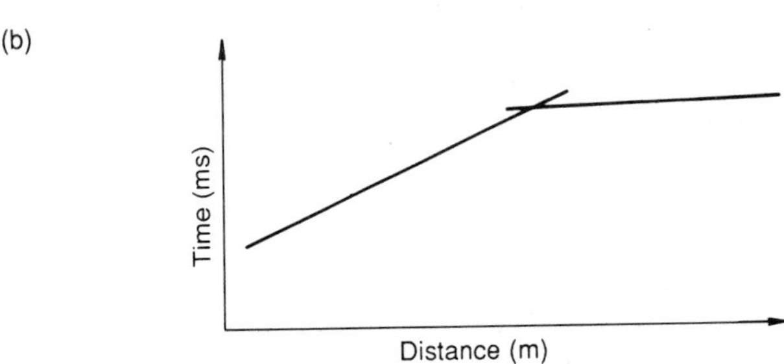

Figure 14.5 (a) A refraction seismogram with the vertical time lines generated in the seismograph. (b) The time–distance plot (see also Figures 1.12 and 14.4)

$$z = \frac{t_0 V_1 V_2}{2(V_2^2 - V_1^2)^{1/2}}$$

Alternatively, the cross-over distance X_c may be used. At the cross-over distance the faster refracted wave overtakes the direct wave and the two travel-times are equal. The direct travel-time is X/V_1 and so

$$\frac{X_c}{V_1} = \frac{X_c}{V_2} + \frac{2z(V_2^2 - V_1^2)^{1/2}}{V_1 V_2}$$

thus

$$X_c = 2z\left(\frac{V_2 + V_1}{V_2 - V_1}\right)^{1/2}$$

and

$$z = \frac{X_c(V_2 - V_1)^{1/2}}{2(V_2 + V_1)^{1/2}}$$

Box 14.1 The travel-time for a refracted wave in the single layer case

The travel-time t is taken over the path ABCD. (Figure 14.3).

$$t = \frac{z}{V_1 \cos i_c} + \frac{x - 2z \tan i_c}{V_2} + \frac{z}{V_1 \cos i_c}$$

$$= \frac{2z}{V_1 \cos i_c} - \frac{2z \sin i_c}{V_2 \cos i_c} + \frac{x}{V_2}$$

$$= \frac{2z(1 - \sin^2 i_c)}{V_1 \cos i_c} + \frac{x}{V_2}, \text{ using } \sin i_c = \frac{V_1}{V_2}$$

$$= \frac{x}{V_2} + \frac{2z \cos i_c}{V_1}$$

$$= \frac{x}{V_2} + \frac{2z[1 - (V_1/V_2)^2]^{1/2}}{V_1}$$

$$= \frac{x}{V_2} + \frac{2z(V_2^2 - V_1^2)^{1/2}}{V_2 V_1}$$

In practice only the P waves are normally used. To record refractions well from a certain depth, geophone spreads out to a distance of some four to five times the expected depth at least are necessary as the critical distance is three to four times the depth. Often only the first arrivals are measured (to an accuracy of about 1 ms), greatly simplifying the necessary instrumentation and interpretation. Of course, if the velocity in a layer increases with depth the travel-time curve will not be a straight line (see p. 130).

An important point is that the refraction method assumes homogeneous and isotropic layers. Also, if velocities do not increase with depth at any interface the refraction method breaks down. In Figure 14.6, where the second layer has velocity V_2 lower than V_1, refraction is away from the interface, there is no refracted wave along the interface and no segment on the travel-time curve, thus this layer will not be detected and V_2 will not be measured. Similar problems arise if the velocity contrast and/or the layer thickness is too small so that the refraction is not a first arrival (Figure 14.6). This will also make estimates of lower layers (such as layer 3) too thick due to the effect of the 'hidden layer' (Hawkins and Maggs, 1961), unless the refractions are recognized as second arrivals. It is important to calibrate calculated depths with borehole data wherever possible, to check against hidden layers. V_2 could also be measured by acoustic (borehole) logging.

We have considered the simple case of two layers; if there are three layers with velocities $V_1 > V_2 > V_3$, critical refractions occur at the two interfaces (Figure 1.12) and there are three sections of the travel-time curve with slopes giving the three velocities. The thickness of the second layer is given by the following for the case of flat layers;

$$Z_2 = 0.5\left[t_2 - \frac{2Z_1(V_3^2 - V_1^2)^{1/2}}{V_3 V_1}\right] \times \frac{V_3 V_2}{(V_3^2 - V_2^2)^{1/2}}$$

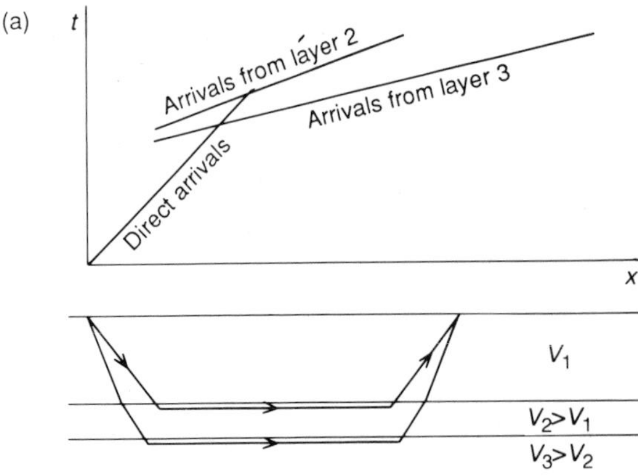

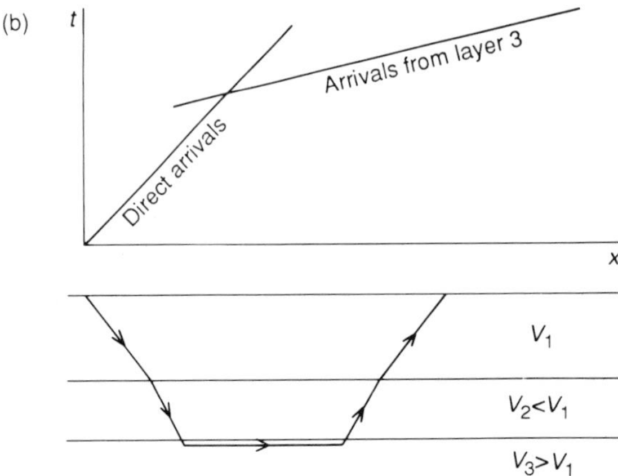

Figure 14.6 The hidden-layer in refraction surveys: (a) a thin layer which does not produce a first arrival; (b) a low-velocity layer that does not produce a critical refraction at all. (After Kearey and Brooks, 1991)

where t_2 is the second time intercept and other symbols are as in Figure 1.12.

For the case of many horizontal layers, all of increasing velocity with depth, the depths can be calculated in succession if a clear refractor can be recognized for each layer. In practice, however, only three layers may be recognizable in shallow surveys. If the interfaces are not horizontal, the dips have to be determined also (see below).

For many thin layers of increasing velocity with depth we have, in the limit, ray and travel-time curves of continually changing slope (Figure 14.7). The simplest case, and a model often used for compacting sediments, is that of a linear increase of velocity with depth z, $V = V_0 + kz$, where V_0 is the surface velocity and k the velocity gradient. Ray paths are then circles with their centres V_0/k above the surface. The radii of the circles is $Z_{\max} + V_0/k$, where $Z_{\max}$ is the maximum depth of the ray, and the travel-time is

$$\frac{2}{k}\sin h^{-1}\frac{(kx)}{(2V_0)} \approx \frac{x}{V_0} - \frac{1k^2x^3}{24V^3}$$

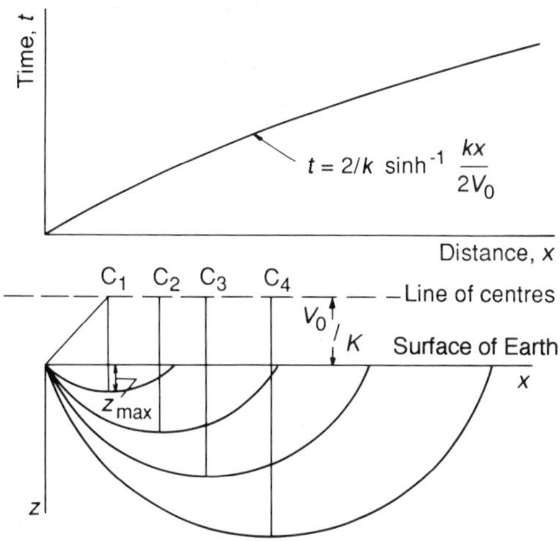

Figure 14.7 Ray paths and a time–distance curve for a linear increase of velocity with depth. (After Dobrin and Savit, 1988)

In sedimentary rocks and velocity gradient k is often about 1 m/s/m. The inverse of the slope at any distance x on a travel-time curve equals the velocity at the maximum depth for rays reaching the surface at that distance x. Other velocity functions have been discussed by Kaufman (1953).

Integral equations relating distance, travel-time and depth of penetration for a ray can be used where the velocity variation with depth is known (or assumed) (see Dobrin and Savit, 1988, p. 467).

So far we have considered a flat Earth, which is accurate for shorter distances, i.e. up to several hundred kilometres. For greater distances, as in teleseismic recording of earthquakes, a spherical or spheroidal Earth model must be used (see Bullen and Bolt, 1985).

14.2.1 Dipping interfaces

An important modification to the above equations is necessary when dip is present, and this must be assumed in accurate work. The critical angle of an incident wave is relative to the normal to the interface (Figure 14.3), and so note that it is at a different angle to the vertical with dip present. Down-dip travel-time (Figure 14.8) is given by

$$\frac{2z_2 \cos i_c}{V_1} + \frac{x}{V_1}\sin(i_c + \theta)$$

where z_2 refers to the depth under the shot, θ is the dip angle and i_c is the critical angle. The measured velocity for V_2 in the second layer is low because of the increasing depth with distance and this is replaced in this equation by $V_1/\sin(i_c + \theta)$. Up-dip travel-time has the same expression with a minus sign in the second term $(i_c - \theta)$. The measured velocity is now high because the depth decreases with

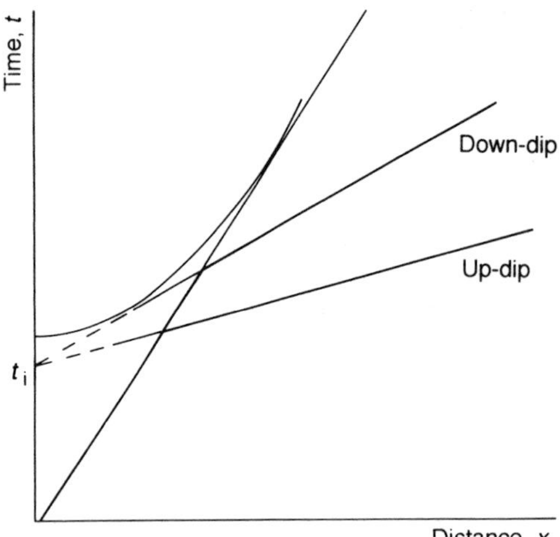

Figure 14.8 The effect of a dipping refractor on the time–distance plot. Note that the intercept on the time axis is not altered. (After Garland, 1979)

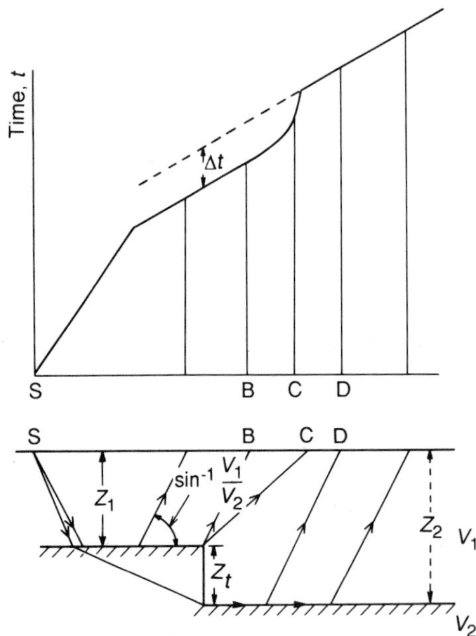

Figure 14.9 The effect of a fault on the refraction time–distance curve. Δt is a measure of the throw of the fault. (After Dobrin and Savit, 1988)

distance. So the refractor is dipping towards the shot-point with the largest intercept time.

The true velocity V_2 is often found accurately enough by using geophone spreads either side of the shot and averaging the measured up-dip and down-dip values. A better approximation is $1/V_2 = 1/2(1/V_d + 1/V_u)$, where V_d is measured down-dip and V_u is measured up-dip. V_2 can be obtained from $\sin i_c$ (see the following) as $V_2 = V_1/(\sin i_c)$.

Since $(i_c + \theta) = \sin V_1/V_d$, and $(i_c - \theta) = \sin V_1/V_u$, both the dip angle and i_c can be found.

$$\text{angle } \theta = 1/2(\sin^{-1} V_1/V_d - \sin^{-1} V_1/V_u)$$
$$\text{angle } i_c = 1/2(\sin^{-1} V_1/V_d + \sin^{-1} V_1/V_u)$$

Depths can then be found from the time intercepts and the known value for i_c. Note that up-dip and down-dip shooting from one shot-point give the same time intercept $t_i = (2z \cos i_c)/V_1$ (Figure 14.8).

14.2.2 Refraction across a fault

Figure 14.9 shows the effect of a fault in the lower bed. The refraction appears as two parallel curves from the upper and lower sides of the fault. In between is a curve produced by diffractions from the fault. The vertical throw of the fault,

$$\frac{\Delta t V_1 V_2}{(V_2^2 - V_1^2)^{1/2}}$$

can be estimated from the difference in intercept times Δt, and the fault located approximately by the change in slope (actually displaced to a somewhat greater distance).

14.2.3 Time–depth (reciprocal) methods

The refraction method so far described produces depths and velocities averaged over the distance of the refracted arrivals used. Better resolution and an approach to spot depths, as needed in engineering geology, for example, can be obtained using time–depth (reciprocal) methods (Hawkins and Maggs, 1961; Sjogren, 1984). In Figure 14.10, assuming the refractor is reasonably plane between D and H and with a small dip (10° or less), with shot-points at A and C, the average time delay (sometimes called time-depth) to geophone B is 1/2(ADEB + CHGB – ADHC times).

The average time delay under S is close to

$$h_B \frac{(V_2^2 - V_1^2)^{1/2}}{V_1 V_2}$$

where h_s is the depth perpendicular to the refractor, thus

$$h_B = \frac{(\text{average delay}) V_1 V_2}{(V_2^2 - V_1^2)^{1/2}}$$

The topography of the sub-surface may then be plotted by striking arcs along the traverse and drawing an envelope.

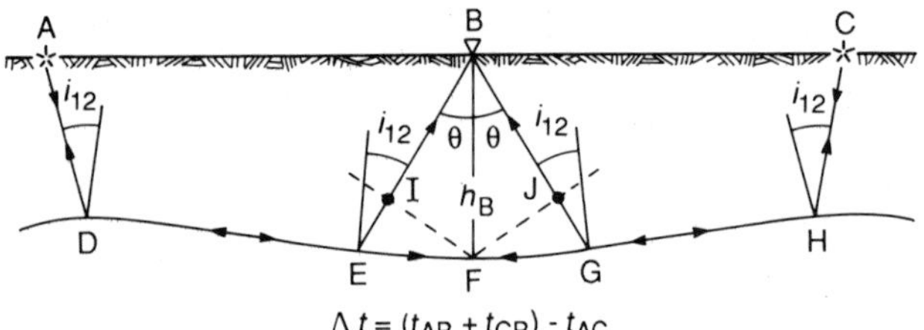

Figure 14.10 Reciprocal methods in refraction. The direct or reverse time (t_{AC} or t_{CA}) is equal to the sum of times A to I and C to J. Thus Δt is the time for the refraction I to B and J to B. (After Robinson and Coruh, 1988)

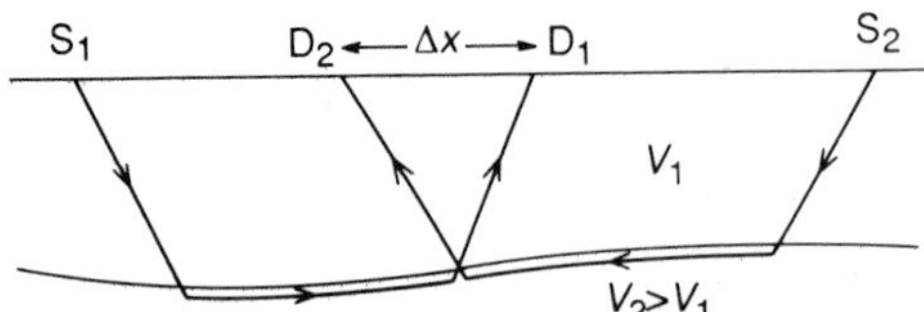

Figure 14.11 The generalized reciprocal method in refraction surveys. (After Palmer, 1990)

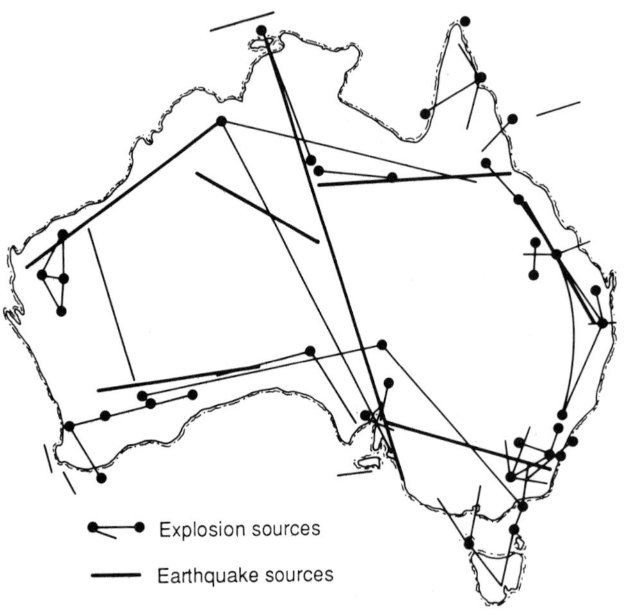

Figure 14.12 Crust and mantle refraction survey lines in Australia. (After Clark and Cook, 1983)

14.2.4 The generalized reciprocal method

This method, first proposed by Palmer (1990), can improve accuracy at pronounced refractor irregularities by using times to two geophones separated so that rays from forward and reverse shots emerge from near the same point on the refractor (Figure 14.11). Only the refractor velocity V_2 is required, not V_1 (Hatherly and Neville, 1986).

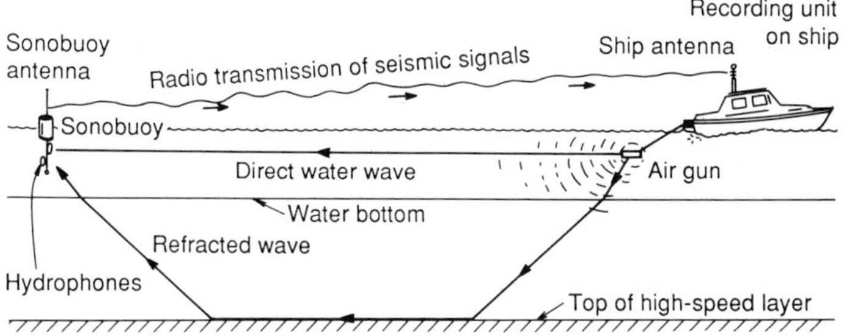

Figure 14.13 Refraction surveying at sea with one ship and the use of a sonobuoy. (After Dobrin and Savit, 1988)

14.3 REFRACTION IN CRUST AND UPPER MANTLE STUDIES

Earthquakes were used as sources in early refraction studies of the crust and upper mantle and are still useful when accurately located and timed. Large explosive sources (tonnes in size, occasionally quarry blasts) are commonly used with lines of seismographs hundreds of kilometres long to reach depths of 30–40 km and more (Figure 14.12). Seismometers or geophones may be placed 1–5 km apart. Nuclear blasts have also been employed (Carder *et al.*, 1966).

In marine crustal studies arrays of pressure-sensitive hydrophones are used to detect the P waves (no S through the ocean). As in all marine geophysics, seismic survey at sea is more efficient than on land. Refraction spreads can be much shorter (say 60 km) because of the shallower crust. In a one-ship survey the ship steams away firing many shots (or an air-gun array) at increasing distances. The recordings can be radioed back to the ship from buoys suspending the hydrophones (Figure 14.13), or recorded on another ship. For such long-distance refraction data, travel-times are often plotted with a 'reduced' time $t - x/v$ where v is an appropriate value (e.g. Figure 10.4). This makes a much more compact plot.

14.4 FIELD PROCEDURES IN REFRACTION SURVEYS

The most common practice in refraction surveying is to record along profiles, preferably along strike direction initially (e.g. see Langston, 1990). Geophones (see p. 141) are laid out from each shot-point to a distance governed by the expected depth of the refractors of interest, which may have been determined in a reconnaissance survey. They are connected to the recorder by special cable. For a shallow survey at an engineering site a simple 12-geophone 'spread' with about 2-m geophone separation may be used. Burial of the geophones reduces the noise level significantly. For wind speeds over about 30 km/h pressure fluctuations and tree and grass vibrations may necessitate

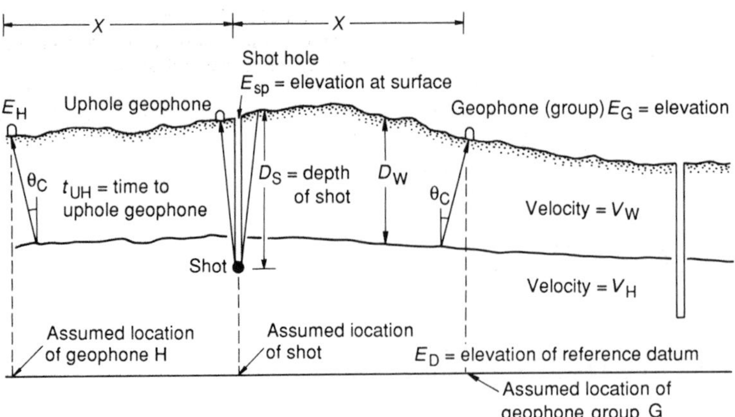

Figure 14.14 Determining static corrections (for height and weathering) from refraction and up-hole times. Data are reduced to a datum level. (After Sheriff, 1984)

suspension of a survey. Other noise sources are traffic, aircraft and 50 or 60 Hz, etc., electromagnetic pick-up from power-lines. Distances to the geophones must be surveyed. In marine surveys distances can be obtained using the P (sound) wave arrival through the water layer as the velocity can be found from the water temperature and salinity.

In refraction surveys explosives or hammer sources are used, although the latter have poor penetration. Shot positions are shifted to a series of positions along the profile and the geophone spreads set to record in both directions from each shot. Note that the interchange of any shot and receiver would give the same travel-time (reciprocity). There are several particular ways of setting out the survey, depending on the geological problem, the number of layers and the amount of dip present. Dobrin and Savit (1988, p. 479) summarize some of these, including broadside shooting across strike.

14.5 CORRECTIONS TO REFRACTION DATA

As in reflection surveys, for accurate work the heights of the geophones and shot-point must be known, usually relative to some adopted datum determined by surveying (Figure 14.14). Special small-scale refraction surveys with close spreads may also be necessary to determine the velocity and depths and thus the time delays produced by the weathered material near the surface which has a very low velocity. This velocity may be only a third of that of the material below and so cause three times the delay of an equal thickness of unweathered rock. Up-hole times at shot-points can be useful (Figure 14.14), with interpolations between holes. The combined height and weathering corrections ('statics') are then added or subtracted from the travel-times (see also p. 161 and Sheriff and Geldart, 1982).

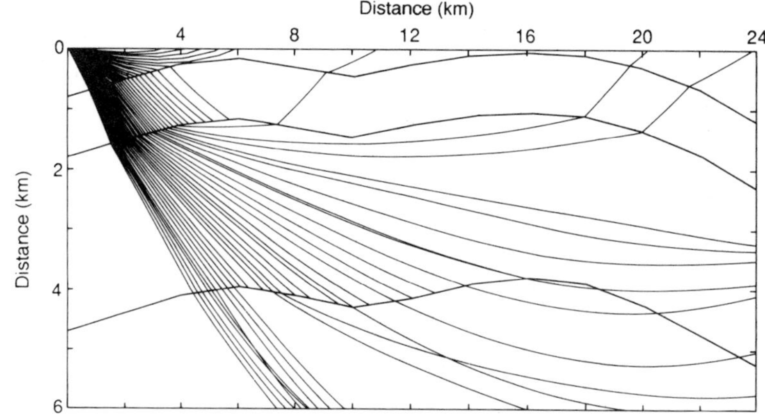

Figure 14.15 Modelling of more complex areas may be carried out by tracing rays through a series of models until the model travel-times match the measured times reasonably well. (After Bayerly and Brooks, 1980)

14.6 REFRACTION AMPLITUDES AND SYNTHETIC SEISMOGRAMS

So far we have only discussed travel-times and distances, but more information is available in the amplitudes, governed by the Zoeppritz equations (Appendix 2) and attenuation rates, and also from the waveforms. The amplitudes of body waves decrease with distance x as $1/x$ (where the refraction is continuous). Refracted or head waves decay more quickly, as $1/x^2$ at distances much greater than the critical distance.

More detail about structure can be obtained by comparing records with synthetic seismograms computed from geological models, but much greater time and effort is necessary. In this method layer thicknesses, dips and velocities are varied until an optimum match is obtained with times, main amplitudes and, to some extent, waveforms (Kennett, 1977; Kennett and Harding, 1985; Vogel *et al.*, 1988). This also involves the modelling of the energy source, transmission effects (including attenuation) and the responses of the geophones, amplifiers and recorders. Inversion techniques have several advantages (Zelt and Smith, 1992). Complex geology may be modelled by ray tracing in which rays are traced through a series of blocks of constant assumed velocity in a model using Snell's Law (Cerveny *et al.*, 1977; Figure 14.15).

Chapter 15
Seismic reflection surveys

15.1 INTRODUCTION

Reflections are produced at any sharp change in seismic (acoustic) impedance ρv (see p. 27), including a drop in velocity (in contrast to the refraction method), and even for a change in density but not in velocity. However, most reflections are related to an increase in velocity and occur at formation boundaries and unconformities but also from within formations. Constructive interference of reflections from alternating shales and siltstones, for example, can produce strong arrivals, whereas thick carbonate sequences are generally transparent (Christensen and Szymanski, 1991).

A survey is often done by a contract company. The seismic crew size can range from 10 to 100 or so in oil or gas surveys, and even 1000 in jungle conditions. Reflection data have greatly improved since the 1950s, with new methods (e.g. common mid-point, p. 147, and three-dimensional (3D) surveys, p. 148), new sources (e.g. Vibroseis-type on land and air-guns at sea), multiple digital recording and computer processing. Most of the increase in data quality is because of the great increase in data quantity with short sampling intervals in space and time.

In Figure 15.1 the travel-time t for a reflection is $(\mathrm{AR} + \mathrm{RB})/V$ for a constant velocity. Thus

$$V^2t^2 = x^2 + 4z^2 \qquad \text{or} \qquad \frac{V^2t^2}{4z^2} - \frac{x^2}{4z^2} = 1$$

i.e. the equation of a hyperbola where x and z are the distance and depth respectively. Thus a plot of measured t versus x, for uncorrected reflection data, produces a hyperbola (Figure 15.2). Figure 15.3 shows an actual reflection record. The effect of the different distances of the geophones is removed by the move-out correction (Figure 15.4 and p. 159).

To get good structural information from seismic data requires good spatial resolution and a good signal-to-noise ratio. Common reflection frequencies and wavelengths are about 30 Hz and 100 m

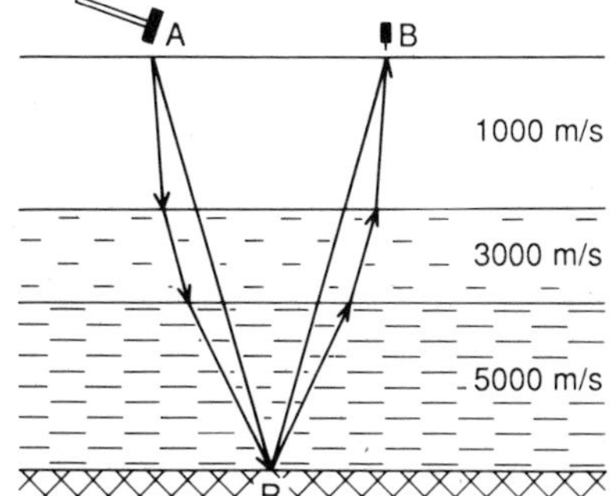

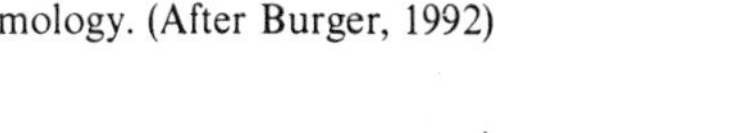

Figure 15.1 Geometry of reflection seismology. (After Burger, 1992)

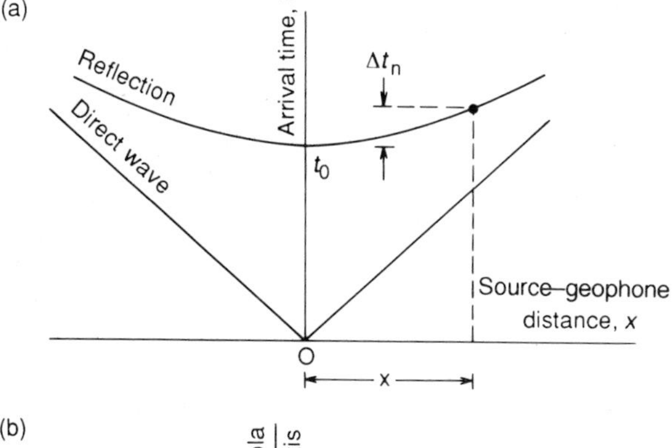

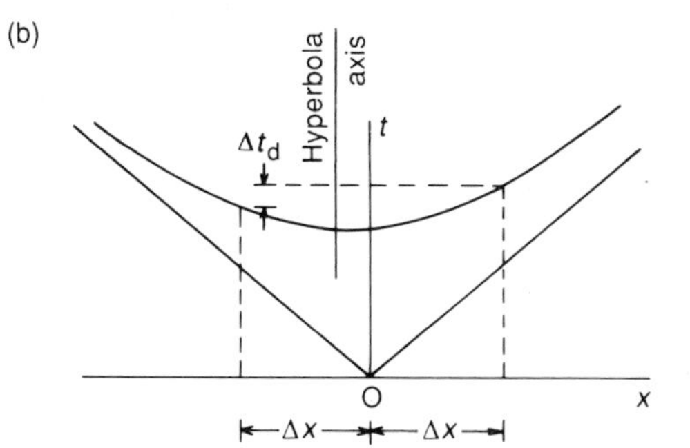

Figure 15.2 (a) Plotting the direct and reflected waves either side of the shot-point and showing the parabolic form of the reflected time–distance curve (for fixed velocity). (b) The effect of dip on the reflection curve. Δt_n is the normal move-out and Δt_d dip move-out. (After Sheriff, 1980)

respectively, so that the resolution, which depends strongly on wavelength, is not ideal (Krapp, 1990). Reflection surveys can delineate structures down to 4000 m depth or so, with a vertical resolution of a few tens of metres. Deeper than this, loss of amplitude of reflections occurs, mainly by geometrical spreading, but also by absorption in the rock, particularly for greater depths and frequencies. One-quarter wavelength ($\lambda/4$) is about the limit of vertical resolution, and the recorded wavelengths increase with greater depth because of the greater absorption of the higher frequencies (the shorter wavelengths). Thus the vertical resolution decreases as the depth of the reflections increases (Figure 15.5). Horizontal resolution is also important and is governed by the Fresnel zone, the area from which

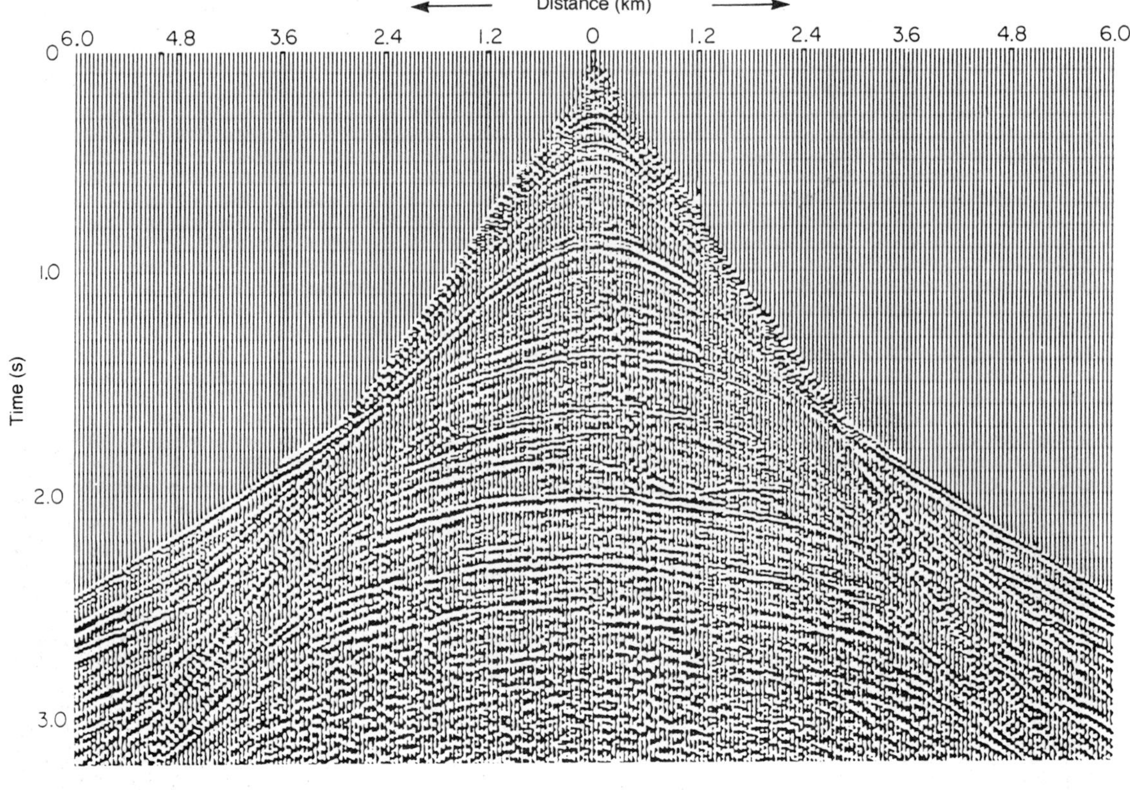

Figure 15.3 A seismic reflection record. (Courtesy of Edwin S. Robinson and Schlumberger Geco-Prakla)

Ray path
V_0
V_1
Datum
Reflector
Before static and NMO correction
After static correction
After both static and NMO corrections
Seismic traces
Two-way time

Figure 15.4 The effect of the statics and normal move-out (NMO) and their removal from a CMP gather. (After Al-Sadi, 1980)

the reflection is in phase and constructive (see p. 190); again this depends on the wavelength.

Reflections are commonly weak in amplitude and never arrive first, so improvement in the signal-to-noise (S/N) ratio is of great importance. Noise includes surface waves, multiple reflections, shallow refractions, reflections which have converted to S, and of course microseisms and

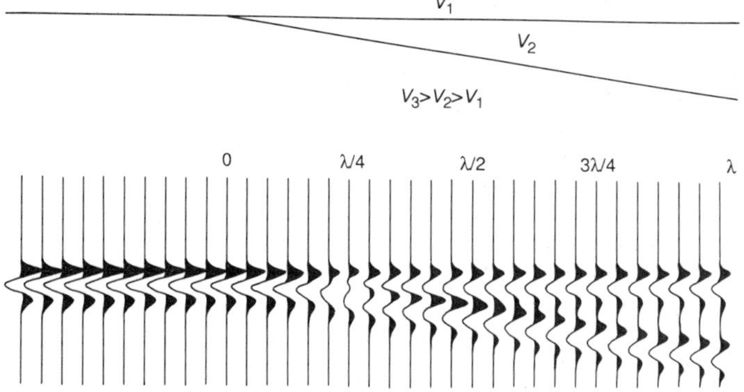

Figure 15.5 Vertical resolution. The effect on a reflection of the increasing thickness of the wedge (V_2). The thickness of the wedge is shown in fractions of the wavelength λ. (After Sheriff, 1980)

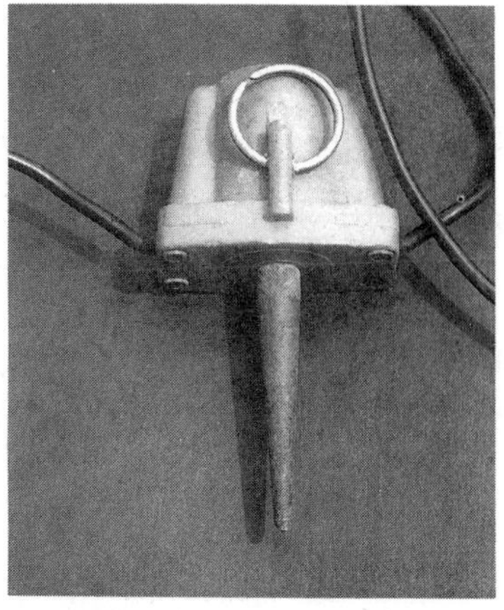

Figure 15.6 Vertical geophones; they are simply pushed into the ground along a traverse and connected by the special connecting cable to amplifiers and recorder

wind and traffic noise. Methods to reduce noise include deconvolution (p. 166) and the common-depth point method (p. 147). The latter is used as an important source of extra (redundant) data allowing summing of reflections.

15.2 INSTRUMENTATION IN SEISMIC SURVEYS

The seismometer (detector) used in geophysical exploration on land is the geophone (Figures 3.1 and 15.6), a small rugged, normally vertical component transducer which is just pushed into the ground or sometimes buried in a shallow hole to reduce wind-generated noise. Geophones sensitive to vertical motion are important in reflection surveys as P-wave reflections are close to vertical in their ground motion. There are also special geophones sensitive to horizontal vibrations for use in S-wave studies.

Geophones are usually the moving coil type, the coil being suspended by a spring and surrounding a permanent magnet which is attached to the casing (Figure 3.1). The relative motion of the coil and magnet produces a small voltage proportional to the velocity. A special cable has positions for easy connection of the geophones to the amplifiers. The geophones are commonly in 96 groups or more in land reflection surveys for hydrocarbons, but only 6 or 12 single geophones are used in small-scale refraction or reflection for engineering geology. In seismic surveys over water, hydrophones are used instead of geophones. Hydrophones are small ceramic piezoelectric devices which produce small voltages from the pressure variations of the P waves in the water (see p. 155).

The natural frequency f_0 of a geophone is governed by the mass m of the moving coil and the elastic constant k for the spring,

$$f_0 = \frac{(k/m)^{1/2}}{2\pi}$$

The natural frequency of the geophones is chosen to be below that of the seismic signals, thus making the geophone response almost flat (uniform) above the resonant frequency (Figure 3.3) (Evenden and Stone, 1971). For reflection surveys, somewhat higher frequency geophones are used (4–15 Hz) than for refraction (1–10 Hz). Reflections are usually in the 20–50 Hz range and have only a few cycles. Refractions have lower frequencies but more cycles. Damping of about 70% is included to remove the resonant oscillations of the geophones. The amount of damping can be adjusted by altering a shunt (resistor) connected across the geophone (Figure 3.1).

Geophone or hydrophone outputs range from tenths of a volt early in the record, from shallow reflections and surface waves, down to about only one microvolt late in the record, from deep reflections. The difference can amount to ratios of 10^4–10^6 times. Hence a large 'dynamic range' and high gain is necessary in the amplifiers (10^5 dynamic range = 100 db) as well as in geophone response in reflection surveys.

Box 15.1 Decibel values and amplitude ratios

Amplitude and power ratios are often expressed as decibels, one-tenth of a bel. The number of bels = $\log_{10}$ (ratio), and the number of decibels = $10 \log_{10}$ (ratio) for power, or $20 \log_{10}$ (ratio) for amplitude, since power = amplitude^2, e.g. 4/1 in power = 2/1 in amplitude; 4/1 = 0.6 bels = 6 db.

db values	Amplitude ratios
6	2 ×
12	4 ×
18	8 ×
20	10 ×
24	16 ×
30	31.6 ×

continued

continued

36	63 ×
40	100 ×
42	126 ×
48	251 ×
60	1 000 ×
80	10 000 ×
100	100 000 ×

Modern recorders are digital, i.e. they include analogue-to-digital (A–D) converters (Figure 15.7) and operate on a series of numbers instead of voltage values as in analogue systems. Digital recording has produced many advances in quality (Hatton *et al.*, 1986). The digital values are recorded magnetically and, amongst other advantages, have a larger dynamic range (about 100 db) and enable computer data processing, such as digital filtering, superior to analogue (electrical) filtering. Analogue amplifiers record on paper or on magnetic tape, with dynamic ranges of only 20 db for paper records and 40 db for analogue tapes.

High-cut filters are sometimes used between the geophones and the amplifiers to reduce high-frequency noise, e.g. above 40 Hz or so in reflection work, and a low-cut filter for ground roll (surface waves). The digital circuitry includes high-speed switches (multiplexers) to feed the incoming data into one continuous stream by sampling the output voltage from each geophone group in turn for about 1 μs each at 2 ms intervals, for example (Figure 15.8). Multiplexing allows the use of only two wires instead of many to connect the geophones to the circuitry. The multiplexers are followed by pre-amplifiers (Figure 15.8), plus an alias filter to cut off frequencies above half the sampling frequency (the Nyquist frequency). This is necessary because sampling the data at intervals (instead of continuously) introduces false frequencies (aliasing) into the data (Figure 15.9). These false frequencies range from half the sampling frequency down to zero. Thus to make sure the frequency range required is obtained, the sampling frequency must be at least twice the required upper value, and the alias filter must cut off those above half the sampling frequency, which otherwise produce false alias frequencies.

Floating-point amplifiers are used in digital recording. These

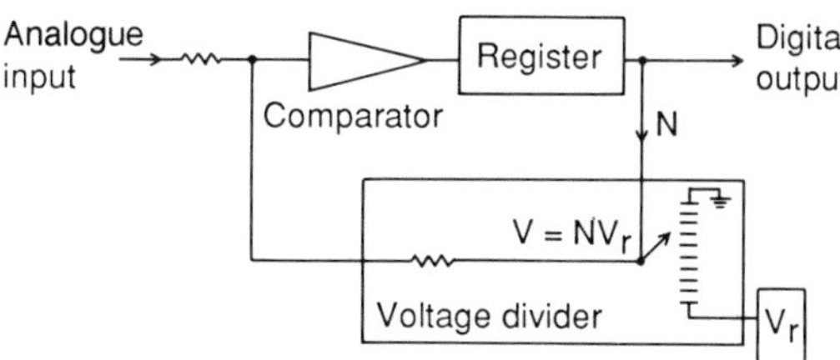

Figure 15.7 The principle of the analogue-to-digital converter (A–D). The input from the geophone or hydrophone group is electronically compared with various values in the voltage divider until a match is obtained and a digital value then put out. (After Dobrin and Savit, 1988)

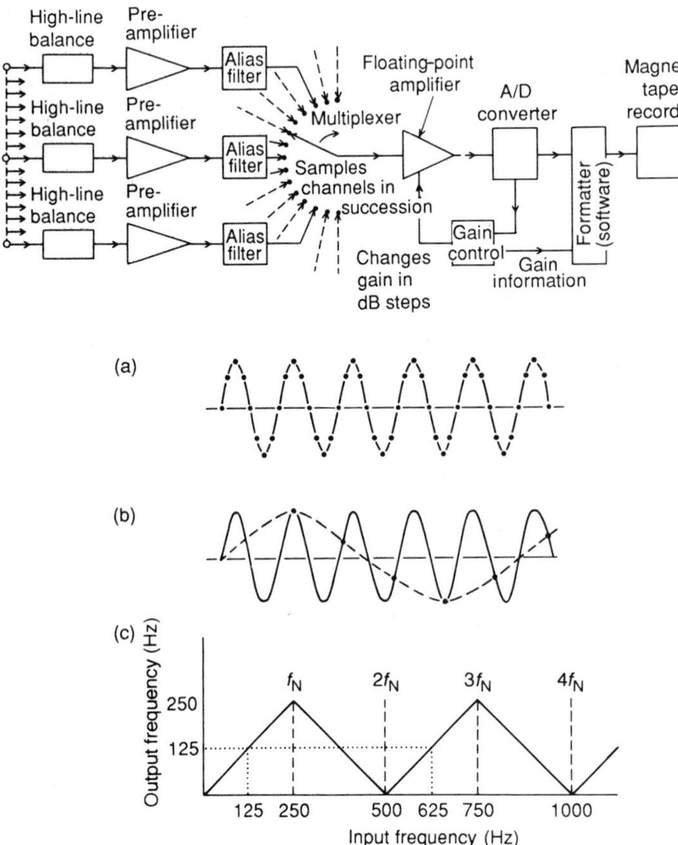

Figure 15.8 A schematic diagram of a multichannel seismic recording system. Note the multiplexer which samples the inputs from the various channels (geophone groups) in turn. Also note the gain control which varies the gain of the amplifier depending on the amplitude of each sample. The gain used is also recorded so that true amplitudes will be known. (After Telford *et al.*, 1990)

Figure 15.9 The effect of sampling discontinuously. The Nyquist frequency is half the sampling frequency and (a) shows a sine wave at less than the Nyquist frequency. (b) A sine wave frequency greater than the Nyquist frequency and the spurious frequency produced by aliasing. (c) The relation between the input and output frequencies with a sampling frequency of 500 Hz (Nyquist frequency $f_N = 250$ Hz). Note that up to f_N the output equals the input frequency, but for input frequencies above f_N spurious lower frequencies are produced. (After Kearey and Brooks, 1991)

amplifiers automatically change the gain (by factors of two at each step) to keep the recorded amplitudes within limits. For example, 15 gain steps of 2 × each give a range of 2^{15} or 90 db. A suitable gain is selected automatically, changing each voltage to a number by comparison of the sample to a series of standard values (Figure 15.8). The gain used at each time is also recorded so that the original amplitudes can be obtained if required, e.g. in looking for 'bright-spots' (see p. 191). The amplifiers usually have a flat response from about 1 to 100 Hz, although higher resolution surveys (for shallow coal, etc.) go up to 300 or 500 Hz. Next the data go to analogue-to-digital (A–D) converters and are demultiplexed back into the various geophone groups and finally plotted. An analogue photographic record (photographing galvanometer light spots) or electrostatic record must, of course, also be made to make the data visible and/or displayed on computer screens.

In analogue tape-recording, frequency modulation is used, or even direct recording in older equipment. Automatic gain (amplification) control (AGC) is used to vary automatically the gain as amplitudes change, i.e. low gain early in the record, increasing during the recording. Negative feedback is used in the amplifiers to achieve this.

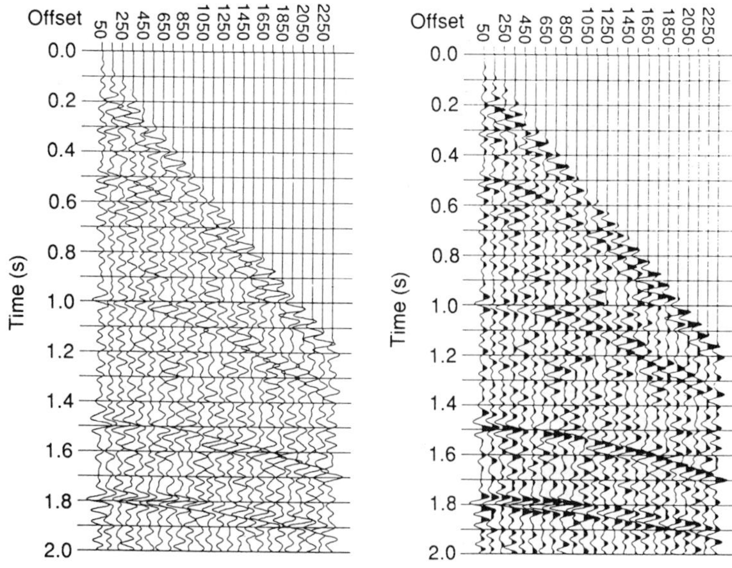

Figure 15.10 Record presentation: (a) wiggly trace, and (b) wiggly trace plus variable width shading. Timing lines are at 0.1 s and offsets (distances) are in metres. (After Robinson and Coruh, 1988)

Otherwise the amplitudes of early reflections would be too large and later arrivals too small to manage.

Photographic or printed records may be variable area, or density, or combinations thereof with wiggly traces (Figure 15.10).

15.3 FIELD METHODS

On land, a survey team first determines the shot and geophone positions, and if explosive sources are used, a drilling truck drills the shot-holes. Geophones are laid out in groups, instead of singly, to help suppress ground roll (surface waves), multiple refractions, wind, traffic or aircraft noise, etc. — all horizontally travelling energy and therefore of low apparent velocity across the geophone array.

Special preliminary 'noise surveys' are often conducted at the beginning of a survey to establish the noise spectrum and the best geophone separation and source type for suppressing noise (Figure 15.11). If the geophones in a group extend over the predominant wavelength of the noise, and the outputs are added, the noise tends to cancel, while the almost vertical and thus in-phase reflections add. The signal-to-noise ratio is then greatly improved in the presence of coherent horizontally travelling noise. The more geophones per wavelength of noise, the better the attenuation of the noise. Another possible source of noise is 50–60 Hz 'pickup' from power lines.

A different type of noise is random (incoherent) noise produced by scattered (diffracted) waves from irregularities, particularly near the shot-point, e.g. boulders, gravel, and changes in the topography, and from tree and grass vibrations, traffic etc. Such incoherent noise is also suppressed by the use of multiple geophones per group if the

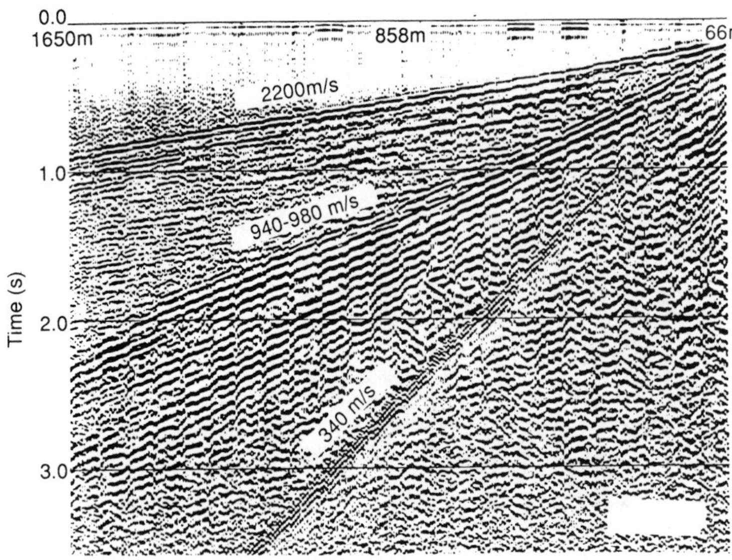

Figure 15.11 A record from a seismic noise survey. Velocities range from 2200 m/s first arrivals to 340 m/s air waves. The 940–980 m/s events are probably Rayleigh surface waves. (Courtesy of Western Geophysical)

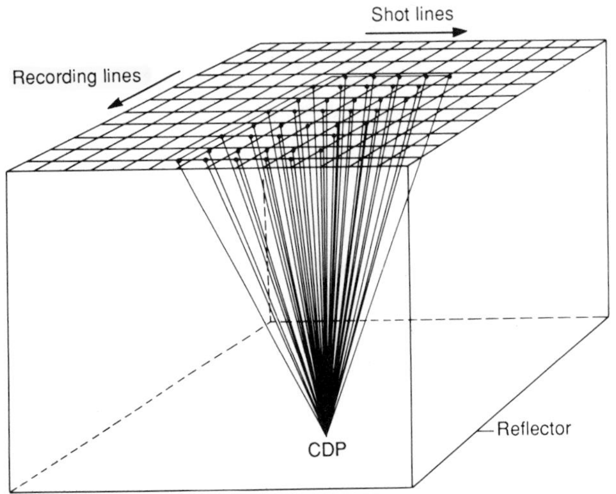

Figure 15.12 Ray paths in a 3D CMP (or CDP) survey

geophones are spread significantly relative to the incoherent noise wavelengths, the effect being proportional to $n^{1/2}$, where n is the number per group.

Many geophones (and hydrophones in marine work) are thus grouped, connected electrically in series or parallel, with perhaps as many as a few hundred per group. The number of groups (and therefore recording channels) commonly varies from 24 to 256, often 96 groups in big reflection surveys, or even 1000 in 3D surveys (p. 148). With a big 'spread' of geophones, the recording truck is connected through a 'roll-along switch' to electrically place the truck at various positions along the traverse. The truck may actually not move all day.

For high-resolution land data, small group intervals, even with geophone arrays equal to the group interval, are sometimes used. A large number of geophones and groups is costly but gives better resolution, and is particularly valuable in sign-bit summing in which only the signs, and not the amplitudes, are recorded.

A special case is the exploration for coal seams and the location of faults which can be very expensive in long-wall mining. The smaller scale and shallow depths require high resolution and higher frequency geophones and recording. Small explosive shots are placed below the weathering. The recording of waves (including surface-type waves) in coal mine seams (in-seam) has been introduced.

15.4 COMMON-MID-POINT (CMP) RECORDING

An important development in reflection surveys was that of common-mid-point recording (Mayne, 1967), also known as the common-depth-point method (CDP), where the reflectors are horizontal. It was introduced in the 1960s and made possible by digital and computer processing (Figures 1.14 and 15.12). Practically all large reflection surveys include CMP recording now, particularly for marine work. The aim is to increase reflection amplitudes but reduce noise by summing many reflections from the same point (strictly small area) with increasing offsets (recording distances). These 'multifold' data are summed after correcting times for normal move-out (the increase of travel-time with distance to the detectors; see p. 159). The seismic noise tends to cancel out whereas the reflections largely add together. The method can be regarded as another form of filtering, and is very useful in helping to remove multiple reflections and other apparent long-wavelength noise.

In CMP recording the records are grouped ('composited') into a 'gather' before adding ('stacking'), one stack for each mid-point. The method gives a lot of additional and redundant data which are also important in determining velocities and dips from the move-outs (see p. 172). CMP gather data are not affected greatly by dip as it comes from a small area on the reflector. Note that CMP summing does destroy some phase and amplitude information, since different reflections are not in phase. This affects higher frequencies particularly. Grouping of geophones has some similar effects.

Today, 96-fold shooting is common, i.e. 96 geophone groups are placed each side of the shot-point, requiring 192 geophone groups. The compositing of records or 'stacking' is done by a field processor or in the playback centre. The length of the geophone spread is called the aperture or spread length and on land may be up to 3 km. 'Vertical stacking' is the addition of records from the same shot-point and geophone, i.e. 'enhancement' by repetition used in Vibroseis-type surveys (p. 149). This is not so easy in marine work where the shot-point is moving, though stacking of near-shots may be done.

Note that CMP processing is inadequate for geologically complex areas. Some of the stacking effects can be corrected by partial migration (p. 186) before stacking, as suggested by a number of authors (Resheff and Kosloff, 1986). (Depth migration of common shot gathers is an alternative to CDP processing and can give correct imaging and better dip and amplitude values according to Resheff and Kosloff.)

15.5 3D SURVEYS

Another important development has been the 'three-dimensional survey' in which a much more detailed investigation is carried out, commonly with geophones placed in a square grid and each position used also as a shot-point (Figure 15.12). This gives a large number of source–receiver pairs and therefore multifold CMP coverage of the area. Line spacing may be as close as 25 m and dual receivers (and sources in marine surveys) may be used. 3D surveys are used where the extra time and expense makes them worthwhile and they are becoming more common, more often for a smaller region such as a production area or prime target (e.g. see Connelly *et al.*, 1991; Vaughan, 1993). The use of dual sources and multiple streamers in marine surveys lowers the cost greatly. 3D surveys give true dips whereas ordinary larger spaced traverses may only give apparent dips of reflectors, unless the survey is in the direction of the dip. Side reflections ('side-swipe') from unknown reflectors can also occur, particularly in 2D surveys.

A 3D survey is costly as 300 channels may be used each time, recording and sampling every 2 ms for 6 s. Then a 1 km survey would produce 36 million samples and with thousands of kilometres of survey, large computers are necessary, sometimes including super-computers (Stanley and Singh, 1991). The Woodside Co recently completed a 2740 km^2 3D survey on the north-west coast of Australia, gathering possibly the largest amount of data collected so far (Vaughan, 1993). Such regional 3D surveys are only justifiable in a known hydrocarbon province but can avoid dry holes or poor placing of wells. The 3D data allow more accurate 3D migration and better amplitude information (Vaughan, 1993).

15.6 ENERGY SOURCES ON LAND

Grouping is also carried out for the sources, i.e. shot-holes or vibrators, often in special patterns to reduce noise. Trial shot patterns may be tested beforehand for new areas. Explosive sources are used in

Figure 15.13 Truck-mounted vibrators used as seismic sources in Vibroseis surveys. (Courtesy of Schlumberger Geco-Prakla)

about half of reflection surveys worldwide (Dobrin and Savit, 1988). A special drilling truck is used to drill the series of shot-holes, which are anything from 1 to 100 m deep. Deeper shot-holes reduce surface wave ('ground-roll') noise which tends to be in the surface weathered layer, perhaps tens of metres deep. Shot depths in the sub-weathered layer or deeper are recommended. The explosive charges, mostly ammonium nitrate (safer than dynamite), are tamped down with soil and water to reduce blow-outs and energy loss to the air, and are detonated with electric detonators. Occasionally, linear sources are used, a rope-like form of explosive (Primacord). Several hundred metre lengths may be ploughed into the ground to a depth of a metre or so. This increases the downward travelling energy relative to horizontally travelling noise.

Conventional explosive recording is generally regarded as producing the best results in most areas and is common for very deep (and crustal) reflections. However, the cost of the explosives, transportation and storage means that vibrator sources (Vibroseis) are cheaper, and coupled with good processing can also give good results (Baeten and Ziolkowski, 1990). Vibration sources are provided by specially designed trucks (Figure 15.13), which are pumped up on an inbuilt large pad about 1 m^2 and set into a vibration by on-board hydraulic pumps or electromagnetic vibrators. The vibration is for about 7–20 s each time, transmitting an extended signal to the ground instead of a pulse (Figure 15.14). Thus a 3 t mass can be vibrated with programmed swept frequency, from low to higher frequency, say 15–90 Hz (Goupillaud, 1976).

The signals from the vibrators picked up by the geophones are cross-correlated with the known source waveform which is transmitted to a computer in the recording truck by cable or radio (Figure 15.14. This cross-correlation is done at various time-shifts, and at a time-shift equal to each reflection time a partial correlation will be

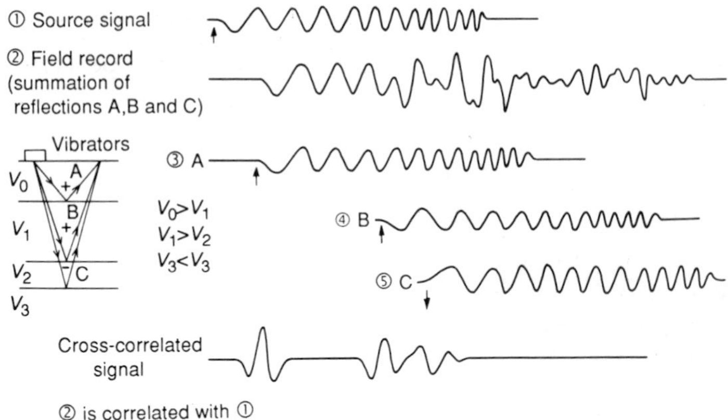

Figure 15.14 Recording and cross-correlation in Vibroseis surveys. The cross-correlation of the recording (2) with the known Vibroseis source signal (1) produces a normal reflection record as from an explosion. (After Dobrin and Savit, 1988)

found, whereas at other times there will be little or no correlation. Such a correlation versus time shift produces an ordinary explosion type record of amplitude versus time.

Several trucks connected by radio signal vibrate simultaneously to produce the 'shot' pattern. Some companies also offer horizontally vibrating shear-wave generators, as S waves have the advantages of determining rock porosity and fluid content. Special S-wave geophones with horizontal coils are then used (Danbom and Domenico, 1986). Vibroseis-type sources do require hard ground, not sandy country, for good penetration and the frequency content is not as high as for explosives.

Surface sources such as Vibroseis have disadvantages in that surface-wave noise from energy trapped in the upper low-velocity layers is much larger than for buried explosives. Advantages are that no shot holes need to be drilled, so it is cheaper and shots can be quickly repeated allowing summation of a number of shots to improve the signal-to-noise ratio. The 'shot' trucks can quickly move on to the next station, as many shot-points can be involved, while the geophone array may be fixed. Also the environment is disturbed minimally; surveys along town streets are possible as the source amplitudes are lower than for explosives.

Another development is the use of a number of earth rammers operated manually in the 'Mini-Sosie' commercial system to give a series of random pulses ($3–8\,s^{-1}$) to the ground (Barbier, 1983). A radio transmitter on the back of each operator sends the time of each impact to the recording truck. The recordings are correlated and summed to give a reasonable signal-to-noise ratio. The method produces high frequencies (40–100 Hz) necessary for very shallow work and is used particularly for surveys for coal, groundwater, fault detection, etc. By summing thousands of values in a few minutes the technique greatly reduces short-duration noise such as traffic. An unusual development in this technique, used in some shallow surveys, is to record only the sign (plus or minus) of the signal and sum them

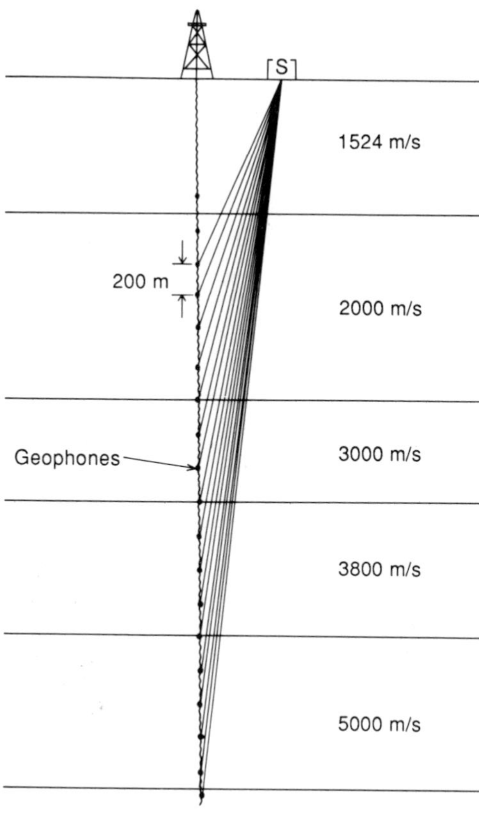

Figure 15.15 VSP (vertical seismic profiling) recording. The shot-point (air-gun) is a short distance from the hole, and geophone positions are down the hole. (After Waters, 1987)

instead of amplitude; this allows the use of many channels, perhaps a thousand.

15.7 VERTICAL SEISMIC PROFILING

This is a fairly recent development in which a down-hole receiver (e.g. a hydrophone) records at a large number of levels 15–30 m apart as it is brought up the hole (Kennett *et al.*, 1980; Hardage and Toksoz, 1983; Balch and Lee, 1984). An explosive or Vibroseis source may be at the top of the borehole, or at increasing distances away from it (Figure 15.15). This provides records of the downward incident waves as well as the upward reflections from beneath a receiver and greatly aids the recognition of the reflectors and their depths. When the receiver is at the reflector depth, the reflection coincides with the incident downgoing wave (Figure 15.16). Also, multiples can be recognized and good synthetic seismograms produced as VSP allows measurement of the downgoing wave-train, plus attenuation and spectral changes better than do sonic logs. The reduced path lengths

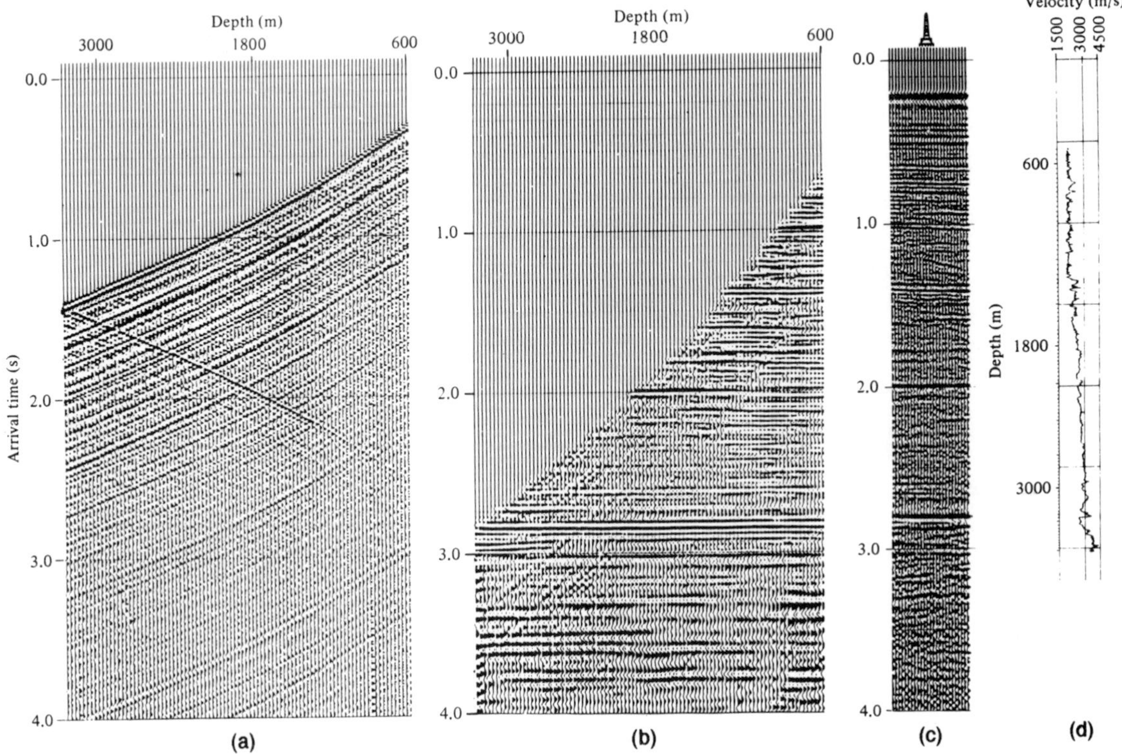

Figure 15.16 (a) A VSP recording. (b) The recording with each trace shifted by the one-way travel-time to the surface, thus aligning reflections (up-going events) horizontally. A shift in the other direction would align down-going events horizontally. (c) A portion of a surface reflection record shot across the well. (Courtesy Sheriff and Geldart, 1982, and Cambridge University Press)

in VSP also mean less attenuation of higher frequencies and better S/N ratios; however, it is expensive.

In VSP data processing the upward travelling reflections are time-shifted by an amount equal to the direct wave arrival time for each trace (Figure 15.16). This makes each reflection appear at the two-way travel-time from the surface, assuming horizontal reflectors and near-vertical paths. This aids correlation with surface-recorded reflections. Deviations are a measure of the dip of the reflectors. When time shifts are made in the opposite sense to reduce first direct arrival data to zero, the downward travelling waves are all aligned, showing which are the multiples.

Another recent innovation is the use of drill-bit energy as a downhole seismic source with surface geophones recording the seismic waves produced by the drill bit (Rector and Marion, 1991). This provides seismic data down the hole without interrupting drilling. Because of strong background noise, the signals are correlated with vibrations coming up the drill-string to aid recognition.

Chapter 16
Seismic surveys at sea

16.1 MARINE SOURCES

Marine surveys take place 24 h/day (if not using explosives) and at a speed of 6 knots or so, hence the rate of data production is very high. This lowers the costs of marine data to about one-third of land data per unit, at least in North America. A 'shot' may be fired each 6–10 s to a minute, or perhaps each 25 m (two hydrophone group intervals). For example, in one survey shots were fired at 8 s (i.e. 25 m) intervals and recorded on 240 channels, producing 67×10^6 samples per km of traverse.

The *air-gun* is the most common source in surveys over water, except for small-scale shallow work when sparkers and some other sources are used (see below). With the air-gun, pulses of air compressed to 2000 or 5000 psi, produced by on-board compressors, are released a couple of hundred metres behind the ship as it steams along (Figure 11.2). Normally an array of air-guns of different sizes is used to produce different frequencies and so a broad spectrum and shorter pulse. Short pulses are required to allow resolution of reflectors.

Other sources include:

(1) *Explosives:* the original energy source in marine work. However, explosives have dropped a lot in popularity as they can only be used in daylight hours, because of safety problems, which lowers the survey speed and raises costs. There is also the problem of bubble oscillations (see below) and the possibility of blowing up the expensive recording streamer, plus environmental concerns.

(2) *Water-gun:* a high-velocity jet of water is pumped into the sea, which produces short pulses and no bubble oscillations (see below) but has less power.

(3) *Vaporchoc:* uses a bubble of high-pressure steam; condensation of the steam in the ocean collapses the bubble quickly.

(4) *Sparker surveys*: involve the use of high-energy (200–5000 J) electrical discharges from condensers into the water. The discharge vaporizes the water, producing a small explosion. Sparkers are

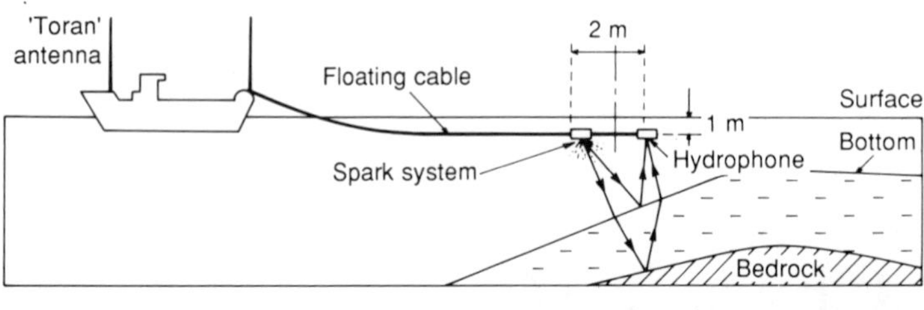

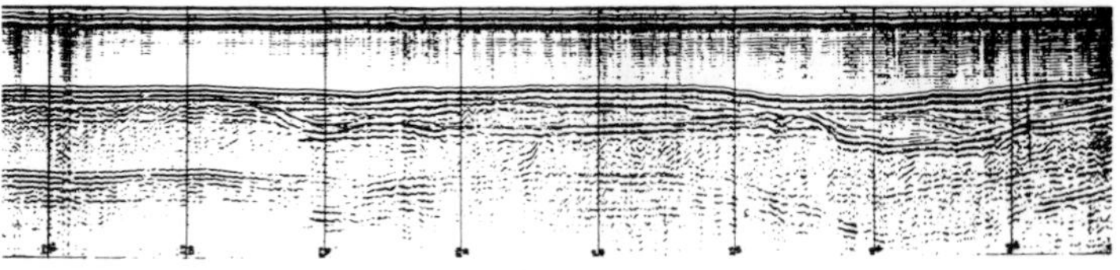

Figure 16.1 A sparker survey in which a high voltage sparker is the source. This is used in shallow harbour or river surveys usually, often with only one hydrophone. Multiples commonly show in the record

mainly used in shallow (about 300 m or less) engineering work, with perhaps only a single hydrophone receiver. The survey is referred to as 'marine profiling' (Figure 16.1). Resolution is good but multiple reflections are strong in sparker surveys and occur shortly after the main reflections in shallow water. Sediments in harbours and rivers are usually thin and so the multiples do not matter too much as the main purpose is usually to map the shallow sediments and bedrock depths; deeper structures are rather difficult to discern. Note that such single-channel surveys cannot give independent velocities.

As with subterranean explosions, as the energy released in marine surveys is increased, the explosion (or underwater air bubble) volume increases and so the peak seismic frequencies fall and wavelengths increase. This increases penetration as the attenuation is less, but the pulse length may also be increased and resolution reduced. The bandwidth (frequency spread) should be adequate as this is necessary for good resolution.

The most efficient depth for the source is at a depth of a wave pressure maximum (antinode), i.e. at a multiple of $\lambda/4$ (one-quarter wavelength), about 10–15 m depth, where the reflected surface signal (ghost; see p. 179) adds to the direct wavelet (there is a pressure node at the surface, zero pressure).

A unique problem of marine surveys is that of oscillations of the bubbles of air or gas produced by the source (air-gun, explosive, sparker). Because of momentum, the bubble released oscillates about its hydrostatic size (i.e. when the gas pressure is hydrostatic) after the initial expansion. It acts like a multiple explosion which confuses the recording (Lavergne, 1970). The frequency of bubble oscillation can be related to the explosion energy by a relation Rayleigh originally developed to explain noises in a kettle.

Bubble oscillations may be removed or reduced by:

(1) exploding at shallow depths less than the radius of the bubble to allow blow-out (an inefficient method);
(2) by recording the oscillation with a hydrophone close to the shot and filtering the known oscillation out of the records, using deconvolution (Backus, 1959; see p. 167);
(3) using an array of different-sized sources (usually air-guns) firing simultaneously; this spreads frequencies over a broad band and causes interference and some cancelling between the bubble oscillations, but not the initial pulses which add together.

16.2 RECORDING AT SEA

Most marine surveys are carried out by one ship towing the source and receivers. The detectors in marine work are small hydrophones, 1 cm in diameter, and comprise piezo-electric crystals of barium titanate or zirconate, or ceramic materials, which produce small voltages as the pressure in the water fluctuates due to P-wave arrivals and other sources. The hydrophones are encased in neoprene plastic tubes (streamers; Figure 11.1), which are filled with oil to give the correct density and flotation. The streamers are controlled by vanes to be at a depth of about one-quarter wavelength 10–15 m, the anti-nodal depth for reflection frequencies, or a multiple thereof. The streamers have a diameter of 5–8 cm and are often of great length (1–4 km), possibly costing over US $1 000 000.

The leading end of the streamer must be weighted to keep the front end down, and a buoy at the end has a DGPS positioner or radar reflector and light to allow monitoring of the tail-buoy position. Also compasses and in-water acoustic networks may be built in to define the orientation of the streamer which affects results.

The hydrophones in the streamers are grouped and connected in reversed pairs so that the pressures produced by seismic waves add, but horizontal accelerations produced by ship and wave motion tend to cancel. There may be 48, 96 or even 480 segments or groups in a streamer, each with 20–32 hydrophones (thus thousands of hydrophones). A recent survey off the north-west Australian coast used three fibre-optic streamers of 320 channels and 5000 hydrophones each, and dual air-guns firing alternatively. Thus six sub-surface lines were acquired, spaced at 25 m. A total of 140 000 km of data was to be acquired at a rate of up to 1000 line km/day (Vaughan, 1993). Multiplicity is necessary for good attenuation of noise which is produced by the ship vibrations, wave motion and the streamer motion through the sea. A reasonably calm sea is desirable. Data from each group are multiplexed and fed back to the recorder on-board in digital format.

Sonobuoys with hydrophones (or phones) suspended beneath them, which radio the hydrophone data back to the ship, are useful for

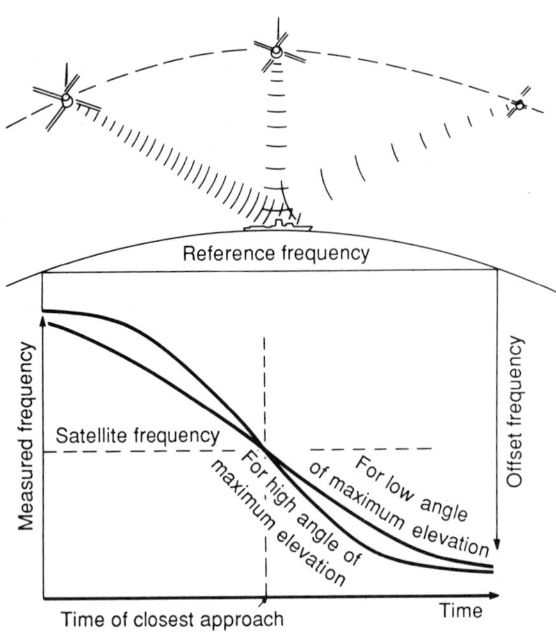

Figure 16.2 Some principles of satellite navigation. For those satellites in polar orbits the time of closest approach gives the latitude, and the rate of change of frequency gives the longitude, the satellite positions always being known. (After Sheriff, 1984)

refraction work where distances are large (Figure 14.13). The travel-time of the water-wave gives the shot distance.

A significant noise problem in marine surveys is that of the multiple reflections between the surface and the sea-floor, particularly with a hard rocky floor; these can swamp the records. This so-called 'ringing' can be reduced by deconvolution (see p. 178). An echo-sounder can be used to continually measure the ocean depths used in the deconvolution calculations.

16.3 NAVIGATION

Navigation in deep-sea surveys is a most important and costly feature. Two or three navigation systems are often run concurrently to allow checking. The American GPS (Global Positioning System) satellite navigation system has become important, particularly well away from shore. There are to be 24 satellites in high orbit so that at least four are above the horizon at any time. The satellites are tracked by US base stations so that their positions are known accurately and broadcast continually by the satellites (Figure 16.2).

The compact GPS receivers are used for geophysical surveys on land and at sea and include computers which automatically tune to the satellites in view and solve for latitude, longitude and height of the receiver. The accuracy for civilian use is about 100 m, but this is greatly improved to a few metres with the use of a fixed auxiliary receiving station and a difference technique.

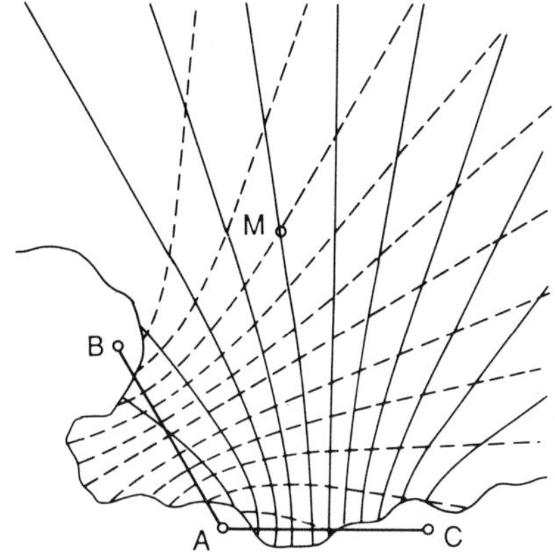

Figure 16.3 Navigation using the phase (or time) differences measured at the ship between signals from three transmitters on shore. Hyperbolae mark the positions of equal phase differences between two transmissions

One way of understanding satellite navigation is the fact that the Doppler shift of the signal can give the time of closest approach, and the slope of the frequency change (Figure 16.2) is a measure of the distance of the sub-satellite point. From the known satellite positions, which are continually tracked, the time of closest approach would give the latitude and the distance the longitude. This method is not actually used for the GPS system; travel-time data from four or more satellites give enough information for the computers to solve for positions.

Near the shore a radar transmitter on the ship can use two or more radar reflector units or Shoran stations which pick up signals, amplify them and re-broadcast them back to the ship, producing continuous position data. Accuracy may be between 5 and 20 m for distances up to 150 km. Being high frequency, line of sight distances are about the limit; however, extended range equipment using refracted and scattered signals from the troposphere (upper atmosphere) can extend this with less accuracy.

There are also continuous-wave systems which do not transmit pulses and use lower frequencies (1–4 mHz), so working at greater distances, perhaps 400 km (see Dobrin and Savit, 1988, p. 145, for the ranges and accuracies of different systems). One method is to use the loci of equal phase or time difference between signals from two shore stations, which are hyperbolae (Figure 16.3). Three transmitters therefore give two intersecting patterns of hyperbolae which give a position, as long as the starting position was known.

Chapter 17
Corrections and processing of reflection data

17.1 INTRODUCTION

Corrections must be applied before interpretation is made, otherwise misinterpretations can easily occur, for example to depths, dips and structures. Data processing (Yilmaz, 1987) is the treatment of the recordings to reduce the effects of noise, distortions introduced by the instruments and by transmission through the Earth (such as attenuation and phase shifts) and it also includes estimation of true amplitude, waveform, frequencies and the velocities in each layer, plus the correct positioning of reflectors (seismic migration). Data processing has become much more successful with the introduction of digital recording and the use of computers, without which much processing would not be possible, and this has resulted in much improved imaging of the sub-surface.

Corrections to reflection data may be summarized as follows:

(1) *Filtering:* as we have seen, low and high cut electrical filters are included in the circuitry to remove or reduce unwanted noise (Figure 17.1). Also special filtering techniques are used to remove remaining horizontally travelling noise, and multiples to some extent, e.g. by using apparent velocity filters (p. 169) or deconvolution. The apparent velocity across the geophone spread for reflections is very high, but for surface and horizontally travelling waves it is very low.

(2) *Normal moveout correction* (NMO) is to correct travel-times for the different geophone distances, ideally to those for vertical reflections (Figure 15.4). It therefore in effect converts reflections to plane waves, at least where the reflector is near horizontal (see below).

(3) *Static-corrections* are used to correct for (i) the differing heights of geophones and the shots, partly due to the topography; and (ii) corrections for the extra travel-time in the weathered layer of soil

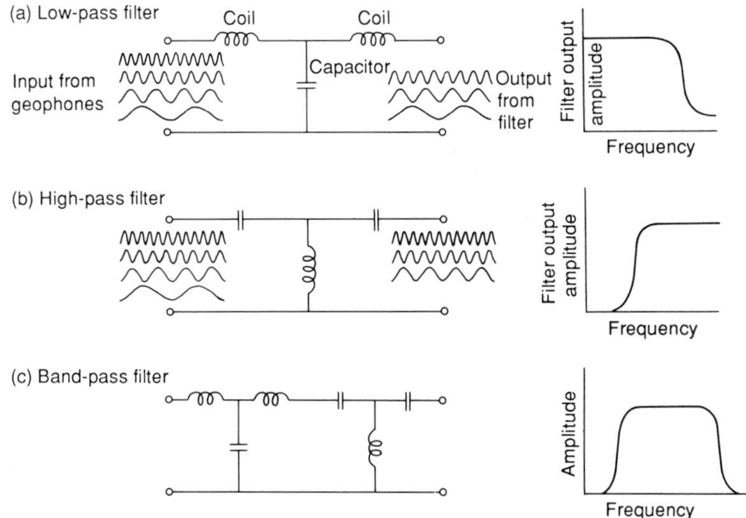

Figure 17.1 Circuits for (a) low-pass, (b) high-pass and (c) band-pass filtering. (After Robinson and Coruh, 1988)

and unconsolidated rock (see Figure 15.4 and p. 161). Unless these corrections are made, quite false maps of the sub-surface would result.

(4) *Deconvolution* is used to help remove multiples (reverberations) and improve the sharpness (higher frequency) of traces (see p. 167).

(5) *Migration* of seismic data is to correct the locations of reflectors (see p. 181). An unmigrated section would be plotted below each shot-point, and so is in the wrong position unless the reflectors are near horizontal.

Velocity determinations are necessary for both NMO and static corrections, for migration and of course for correct depth determinations. It has been described as the most important aspect of data processing (see p. 171).

17.2 MOVE-OUT

This correction converts the time of a trace at, for example, offset x to that at zero offset, i.e. the time that would be recorded at the mid-point (Figure 15.4). It is often called a dynamic correction as it decreases with time down the trace. Applying NMO corrections in CDP data is analogous to light focussing; this seismic focussing removes the time delays caused by the different geophone distances and equalizes travel-times, as does a lens.

As we have seen, the travel-time for a reflection in a simple one-layer case is

$$t = \frac{2[(x/2)^2 + z^2]^{1/2}}{V_a}$$

so

$$t^2 = \frac{x^2 + 4z^2}{V_a^2} = \frac{x^2 + t_0}{V_a^2}$$

Note that a plot of t^2 versus x^2 would therefore give a straight line, but in practice this is not quite so due to higher terms.

Here t_0 is t at $x = 0$, $t_0 = 2z/V_a$, and is t for a vertical reflection (not the t_0 of refraction).

$$t = [1 + (x/V_a t_0)^2]^{1/2}$$

if x is much less than $2z$, then by the binomial expansion

$$t = t_0\{1 + 1/2[x/(V_a t_0)]^2 - 1/8(x/V_a t_0)^4 + \cdots\}$$

The move-out Δt between any two geophones is thus

$$\frac{x_2^2 - x_1^2}{2v^2 t_0}$$

using the first term only.

With one geophone at the shot-point, $x_1 = 0$, and *normal move-out* (NMO)

$$\Delta t \approx \frac{x^2}{2V_a^2 t_0} \quad \text{or} \quad \frac{x^2}{4V_a z}$$

Note that move-out increases as x^2; and inversely as V_a^2 and t_0 or depth z. Also, where velocity increases with depth, root-mean-square velocities V_{rms} should replace V_a in expressions (see p. 173), therefore NMO values can give approximate rms velocities (in practice 'stacking' velocities). This is the basis of modern computer velocity determinations.

NMOs are removed from data to help in the recognition of reflections. It is essential to see the geology and to recognize multiples and diffractions. As we saw they are used to determine velocities. Dipping reflections have some extra move-out (Figure 15.2) which is removed in special dip move-out processing (DMO). Dip move-out causes a tilted line-up of times. The dip is given by

$$\sin\theta \approx \frac{V(\Delta t_d)}{2x}$$

where t_d is the dip move-out and Δt_d is time difference between end geophones (at positions $\pm x$). Other remaining move-outs could be for multiple reflections or diffractions.

After NMO corrections are made, records may be stacked (added), for those of the same travel-times. Fortunately, in CDP stacking of data the effect of dip is small as the reflection groups are from a small area. Wide spreads are needed to stack CDP records, and to determine dips.

17.3 STATIC CORRECTIONS

Corrections for elevation and weathered layer thickness variations are most important in land surveys and are one of the greatest sources of error. These corrections are called static corrections since they do not change with time down the trace as do the move-out corrections.

If elevation differences were not removed, a traverse over a hill would show up in the travel-times as a false syncline. Relative elevations of geophones and shots are 'moved' to a certain datum level using the sub-surface shallow velocity and assuming vertical rays (Figures 14.14 and 15.4). The datum level is commonly taken at or below the base of the weathered layer. It may be sloped in some cases.

The weathering correction is the most difficult to do accurately. The low velocities in the weathered layer, which is often above the water-table, cause significant time delays as the velocities in the weathering can be as low as 300–1000 m/s. Thus time delays may be as much as three to four times that in the same thickness of non-weathered rock and quite variable. The weathering correction must be determined from the early refracted arrivals in the reflection records, or from special weathering surveys and up-hole times. Computer programs are available to automatically pick and process such refractions. Fortunately, reflection paths are near-vertical because of refraction into the low-velocity weathered layer, and corrections can be made by simply adding or subtracting, once the velocity and depth is known. Delay-time (reciprocal) methods (p. 133) are preferable to intercept times.

Up-hole times from deeper shot-holes and wells using shots at several depths may be available to provide valuable information on weathering velocities (Figure 18.2). A plot of those times versus depth is used if enough data are available. Sonic logging down-hole (p. 174 and Figure 17.2), in which a probe about a metre long is used with

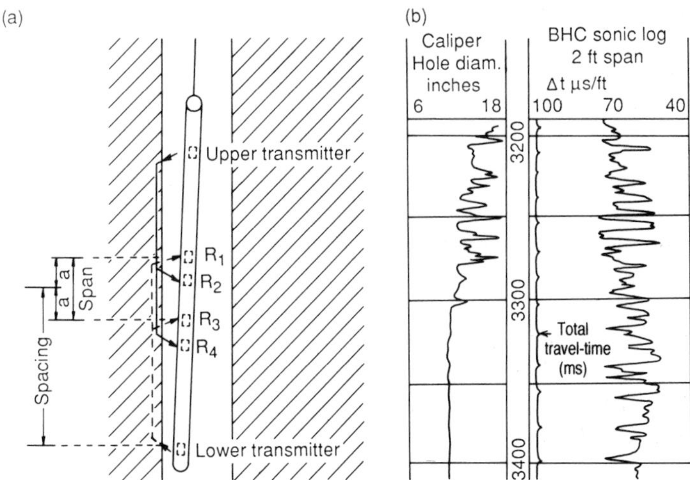

Figure 17.2 (a) A schematic compensated sonic logging sonde. (b) A sonic log. (After Sheriff, 1984)

in-built transmitters and receivers, will provide detailed velocities, where available.

The 'weathering' correction is not normally applied in marine surveys, i.e. sea-floor topography is not corrected for. But hydrophone and shot depths would be corrected to mean sea level, and in some cases corrections may be made for tidal variations.

Static corrections have become even more important with CMP stacking; otherwise stacking of reflections is affected too much, and the pulse shape can be lost. Manual corrections are tedious, so automatic static correction methods have been developed using computers and have been very successful (Hileman *et al.*, 1968). Approximate static corrections, however, may still be made beforehand by manual methods, residuals being used for a final computer determination.

Companies tend to have their own computer programs for automatic or semi-automatic statics correction. The program may be arranged to obtain corrections that best align reflections in a CMP 'gather' so that they are in phase. The redundancy of data of CMP recording is necessary to do this. Each trace from a particular shot can be assumed to have the same shot correction, and similarly each geophone group can be assumed to have a certain static correction. Then a matrix of values is set up in the computer for solving by cross-correlation, etc. A large number of times may have to be averaged to remove errors in static corrections (see Dobrin and Savit, 1988, p. 232).

17.4 DIGITIZATION AND FILTERING

As we saw earlier, in digital recording the continuous output of the seismometers and filters in the form of small voltages is sampled at short time intervals (millisecs) and each sample is recorded as a number in the binary system (a sequence of digits that are either 0 or 1). These numbers can be handled by digital computers for data processing, e.g. further filtering, convolution, amplifying, migration, etc. These operations can then be carried out much more efficiently and with a much greater dynamic range than by the older analogue methods.

The data must be sampled frequently enough to capture all the higher frequencies required and this is satisfied as long as the sampling frequency is at least as high as twice the highest frequency involved (Figure 15.9). Frequencies above twice the sampling frequency appear in the output as spurious lower frequencies (aliasing). Hence the need for anti-alias filters to eliminate such frequencies before digitizing.

The voltages from the geophones or hydrophones may first be electrically filtered to reduce low-frequency surface-wave noise. Electrical frequency filtering can be readily understood in terms of

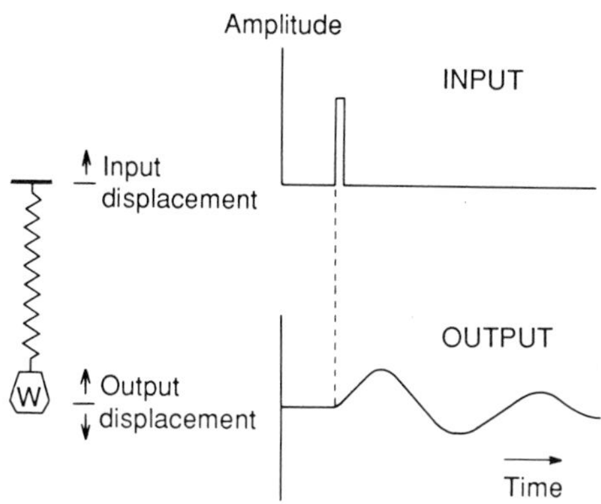

Figure 17.3 Illustration of an impulse response, the output of a system for an impulsive input. This is the analogue in the time domain of frequence response. (After Kearey and Brooks, 1991)

frequency response curves and high- or low-pass filters (Figure 17.1). The data are then multiplexed into a single stream for the analogue-to-digital converter (Figure 15.8).

Digital filtering may be either frequency filtering (more efficient than electrical filtering; Kanasewich, 1990), or inverse filtering. The aim of inverse filtering is to reverse the process and determine the original shape of the reflection wavelet before the filtering effects of the Earth (absorption, multiples, etc.). This could be done by the deconvolution process (p. 167). *f–k* filtering (p. 169) is also available to remove surface waves and other noise. After filtering and amplification, the digital values are demultiplexed back into their original recorder groups and displayed in analogue form.

The *impulse response* is the time analogue of a frequency filter. The impulse response is the output of a system for a very short impulse (spike) input. A simple example is the sudden displacement of a mass suspended by a spring (Figure 17.3). An input wave can be thought of as being made up of a number of spikes, each producing an output whose shape is that of the impulse response, but with amplitude proportional to the input spike and of the same sign (Figure 17.4). The impulse response can be calculated from the frequency response of a device through the Fourier transform.

17.5 THE FOURIER TRANSFORM

The Fourier transform may be used to convert a series of amplitude values versus time into a series versus frequency, an amplitude spectrum, plus a phase spectrum which is also necessary for complete specification of a wave. The phase values are those at the beginning of the wave (Figure 17.5). Thus the transform allows conversion of data from measurements versus time, as in a seismogram, into its spectrum, which is very useful for filtering, noise suppression, etc. The spectra can then be transformed back into the time domain, so that data in the

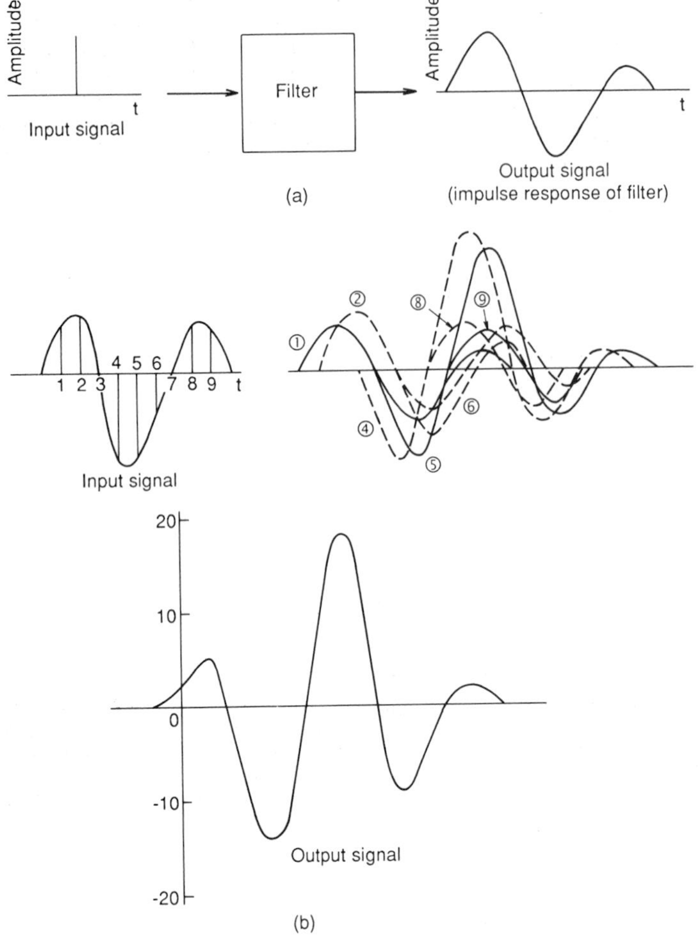

Figure 17.4 The impulse response and its use in determining the effect of a filter (e.g. seismograph, the Earth) on any input. The output curve is the result of superimposing the impulse responses of various amplitudes at the sampled times 1, 2, 3, etc. (After Dobrin and Savit, 1988)

time and frequency domains are interchangeable. Figure 17.6 illustrates some examples.

The transform is based on the Fourier series which expresses any function as a series of sine and cosine waves (Figure 17.5). The Fourier series consists of the fundamental frequency f_0, which is the inverse of the time length of the data, plus a series of sinusoid harmonics of frequency $2f_0$, $3f_0$, $4f_0$, etc. + a constant. The series is described by a series of amplitude and phase values versus frequency (amplitude and phase spectra) or separate sine and cosine transforms. A good example is that a frequency response curve is the Fourier transform of the impulse response.

Mathematically, the Fourier transform is expressed as an integral, a continuous function for an infinitely long recording with the harmonics separated by infinitely small amounts. In practice, data are divided into frequency slices and mean frequencies in each slice are

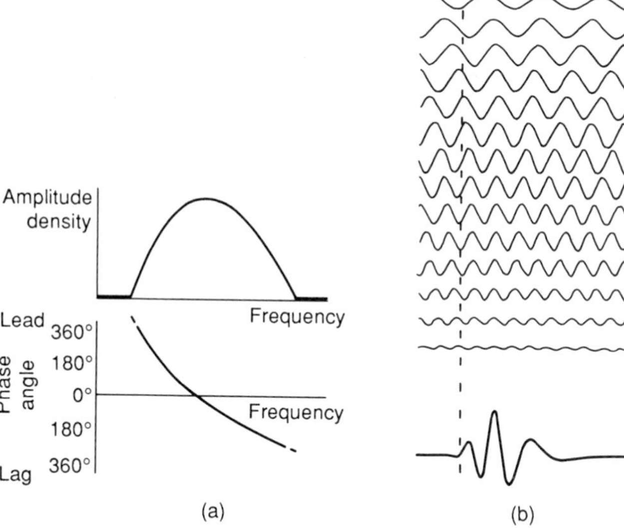

Figure 17.5 The Fourier transform of a waveform: (a) the amplitude and phase spectra; (b) the synthesis of the waveform by the summation of sinusoidal waves at various frequencies with the amplitudes and phases (time shifts) as in the transform spectra. (After Dobrin and Savit, 1988)

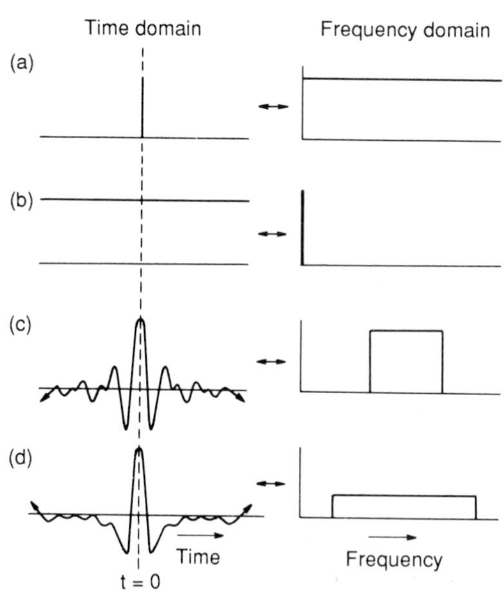

Figure 17.6 Some examples of Fourier transform pairs in the time and frequency domains: (a) a spike or ideal impulse; (b) a bias; and (c) and (d) transient waveforms approximating seismic pulses. (After Kearey and Brooks, 1991)

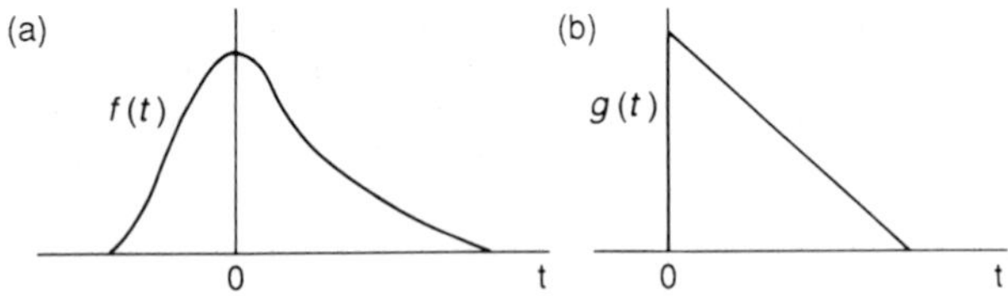

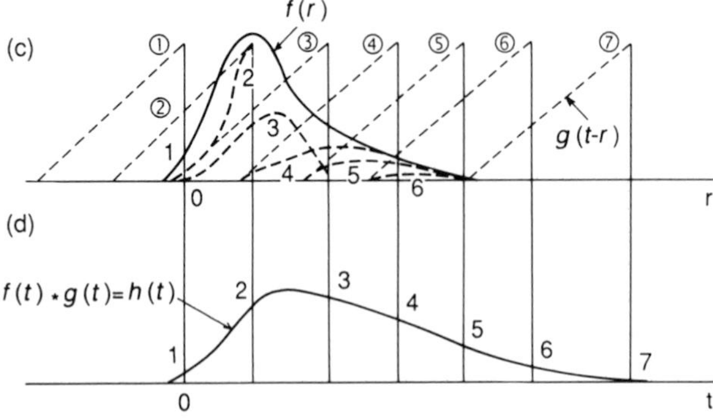

Figure 17.7 The convolution of two functions $f(t)$ and $g(t)$ giving the result $h(t)$. If $f(t)$ is an input signal and $g(t)$ the impulse response of a filter, $h(t)$ is the output. (c) shows the multiplication of the two functions with one reversed and moved past the other so as to have the correct time relation with each other. (After Dobrin and Savit, 1988)

used. Note that if $F(f)$ is the Fourier transform of $f(t)$, $f(t)$ is the transform of $F(f)$, where t is time and f is frequency.

$$F(f) = \int_{-\infty}^{\infty} f(t) \cos 2\pi ft \, dt - i \int_{-\infty}^{\infty} f(t) \sin 2\pi ft \, dt$$

and

$$f(t) = \int_{-\infty}^{\infty} F(f) \cos 2\pi ft \, df + i \int_{-\infty}^{\infty} F(f) \sin 2\pi ft \, df$$

(For more detail, see Yilmaz, 1987, Dobrin and Savit, 1988, Chapter 6, or texts on applied mathematics.)

17.6 CONVOLUTION, DECONVOLUTION AND CORRELATION

Convolution is the change of waveform in passing through a linear filter (Sheriff, 1984). A seismic wavelet from an explosion or other source is filtered and changed by the characteristics of the Earth, geophones, the amplifier and the recorder. The effect of a filter on an input can be calculated by convolving the input function with the impulse response of the filter. In Figure 17.7(c) the second function is reversed, the input slid past the first function and corresponding values multiplied. The reversal of one function is so that the corresponding values are in the right time relation to each other. Note that the input samples, a series of spikes, are replaced by a series of the

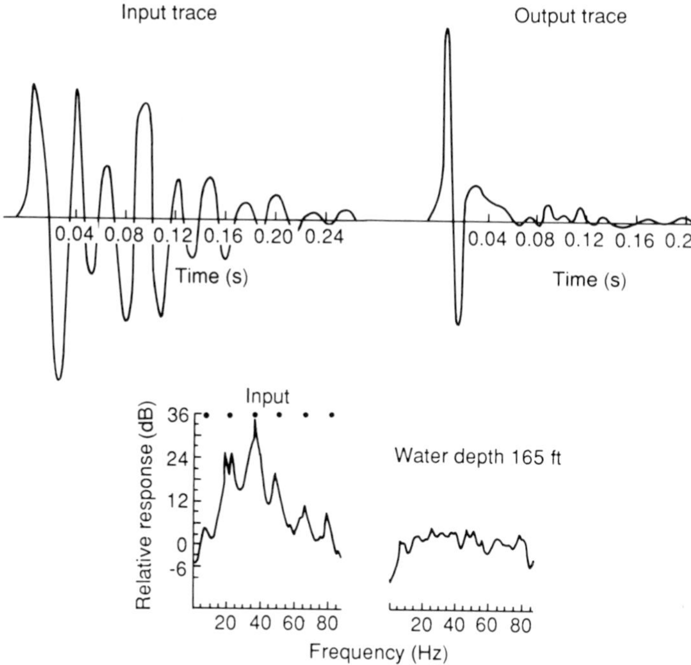

Figure 17.8 Waveforms and spectra of a reflection signal before (input) and after deconvolution (assumed minimum phase). (After Dobrin and Savit, 1988)

impulse responses of the filter of appropriate amplitude.

Mathematically, convolution of two functions is written as

$$c(t) = f(t) * g(t)$$

An important fact is that the Fourier transform of a convolution of two functions is equal to the product of the transforms of the two functions, written as $C(t) = F(t)G(t)$. Thus convolution in time is equivalent to multiplication in frequency, and convolution in frequency is equivalent to multiplication in time. This 'convolution theorem' is often used to calculate Fourier transforms more simply.

Deconvolution is inverse filtering where we have a known output (such as a seismic record) and we wish to determine the shape of the waveform $\{f(t)\}$ before it was affected, say, by the response $g(t)$ of a seismograph or of the Earth (Yilmaz, 1987). Mathematically, this is written as $f(t) = c(t) * 1/g(t)$, or in the frequency domain $F(f) = C(t)/G(t)$.

Actually we do not know exactly the filtering characteristic properties (impulse response) of the ground, so special statistical methods using least squares are employed to estimate the filter properties. An important method is that of Wiener (see Lines and Treitel, 1984). Thus deconvolution is also a kind of filtering and a means of reducing noise and distortions in the data. For example, 'deconvolution before stacking' is designed to make wavelets more impulsive, like a spike, and reflections sharper, i.e. it enhances the higher frequencies, which tend to attenuate more quickly in the Earth than lower frequencies (Figure 17.8). It is also useful in reducing

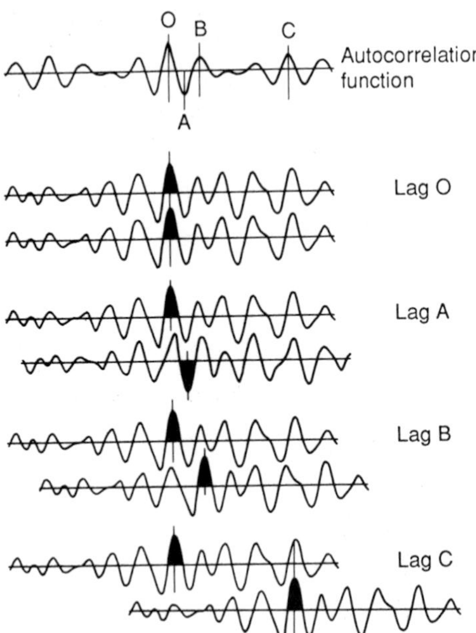

Figure 17.9 Calculating an autocorrelation function by multiplying the waveform by itself at each sample-time and separating the waveform and its copy by increasing amounts (lags). (After Dobrin and Savit, 1988)

multiples as they tend to have lower frequencies. Deconvolution after stacking is to reduce the multiples remaining after stacking.

Autocorrelation is another way of determining the frequency spectrum of a waveform. It is similar to convolution; the waveform is multiplied by itself for various relative time shifts, but without reversing it (Figure 17.9). The sums of the products are plotted against the time shifts to give the autocorrelation function. Maximum autocorrelation values are obtained at time shifts equal to periods in the waveform. So multiples, for which it is particularly useful, and any other unwanted frequency (the inverse of the period) may be identified and later filtered out. Also, the Fourier transform of the resulting autocorrelation function is the square of the amplitude spectrum, i.e. the power spectrum.

Cross-correlation is the same operation, but for two different waveforms or functions. It measures the similarity of the two functions for different time shifts, as in the correlation between the Vibroseis input to the ground and the recorded signals to search for reflection arrival times (p. 149). Maximum correlation is at the time shift equal to a reflection travel-time. It is also used in calculating static corrections. Cross-correlation calculation is thus very similar to convolution but there is no reversal of a waveform. It is equivalent to the multiplication of the amplitude spectra with the subtraction of the phase spectra.

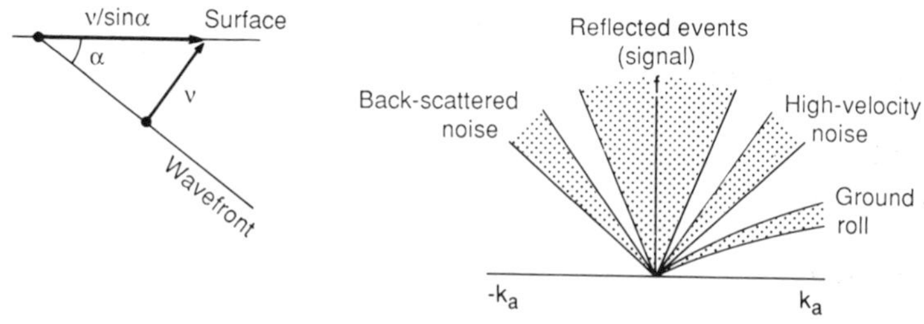

Figure 17.10 (a) The apparent velocity ($v/\sin\alpha$) of a wave-front across a surface geophone spread. (b) The frequency/wave-number (f–k) plot for a seismic gather with reflections and noise. Wave number (k) is the inverse of wavelength ($2\pi/\lambda$)

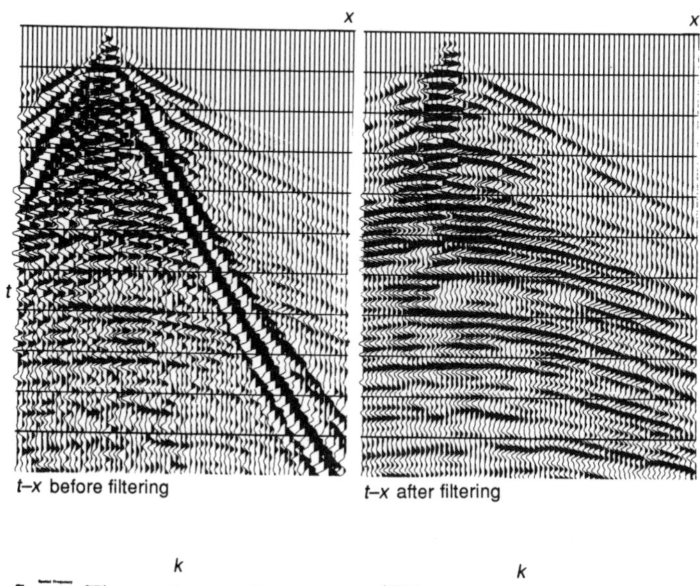

Figure 17.11 An example of a seismic record (t–x) and f–k plots before and after filtering. (Courtesy Seismograph Service Ltd)

17.7 VELOCITY (f–k) FILTERING AND THE τ-p TRANSFORM

Apparent velocity filtering to reduce noise, such as surface waves, uses the fact that different wave types have different apparent velocities across the geophone array. Surface waves travel horizontally across an array, whereas reflections arrive almost vertically and so have a much higher apparent velocity (Figure 17.10). This filtering is often carried out in the frequency/wave-number domain (f–k), transforming the data from time versus distance to frequency versus wave number (k, the inverse of wavelength) (Figures 17.10 and 17.11). Reflections in f–k space tend to map about the vertical axis while the surface waves tend to map closer to the horizontal wave-number axis. However, f–k filtering can cause signal distortion and smoothing.

The τ-p transform (slant stack) is a method of filtering used primarily in filtering before stacking, giving a clearer separation between reflections, surface-wave noise, refractions and diffractions. It also has the advantage that refractions can be represented in the same transform as reflections (Stoffa, 1989; Kappus *et al.*, 1990).

The method transforms data from the usual time–distance plot to a plot of two-way reflection time (at $x = 0$) versus $\mathrm{d}t/\mathrm{d}x$, the horizontal slowness, i.e. the ray parameter $p = \sin i/V$. The transform is a sum of the data from each trace along the slope or slant of $\mathrm{d}t/\mathrm{d}x$. Surface-wave noise (ground roll) transforms into a small area near time zero, and refractions to points at their zero offset, because they have a constant slope $\mathrm{d}t/\mathrm{d}x$. Reflection and diffraction hyperbolae transform into ellipses that do not cross, even if the hyperbolae cross when plotted as time versus offset. The separations allow filtering out of unwanted components while in $\tau - \rho$ space.

Chapter 18
Seismic velocities

18.1 INTRODUCTION

Unlike radar velocities, seismic velocities vary a lot both vertically and, to a lesser extent, horizontally. Ocean velocites vary much less, although there is a low-velocity 'Sofar' layer centred at a depth of about 1200 m in deep ocean (Figure 9.6). Seismic velocities (Table 2.1) vary from about 300 m/s for unconsolidated surface material on land and about 1500 m/s in the ocean, to 5000–6000 m/s in deep sedimentary basins (Christensen, 1989). Velocity variations can affect seismic data in a variety of ways, producing, for example, false anticlines which are not there (Cordier, 1985).

Relations between velocity and density may be found for particular rock types, more accurately if in a restricted region (Figure 18.1). Porosity is, in turn, also involved through the effect on elasticity and density. Porosity is affected in sedimentary rocks by their maximum depth of burial. Note also that there is increasing velocity with formation age, because of the burial and compaction during the history of the formation.

There are a number of defined velocities:

(1) The *apparent velocity* across the geophone spread, which is equal to dx/dt (Figure 17.10). The apparent velocity of surface waves is their true velocity when the geophone array direction is in the direction of wave travel, refracted waves similarly give true velocities if the refracting surface is horizontal. For a truly vertical reflection the apparent velocity equals infinity as all reflections arrive simultaneously, but is finite from a dipping interface, decreasing with dip. The very different apparent velocities for reflections and surface waves makes apparent-velocity filtering very useful (see p. 169).

(2) The *average velocity* (V_a) is the depth of the reflector divided by the one-way reflection time and is used for converting reflection time to depth. For a number of layers,

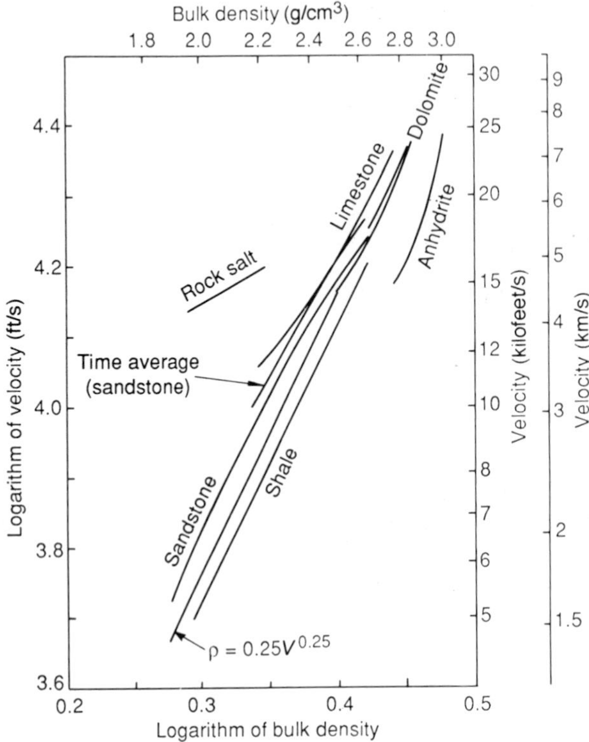

Figure 18.1 The relation between velocity and density in rocks of different lithology, showing both Gardner's and the time-average relations. The time-average equation is $1/V = \varphi/V_f + (1-\varphi)V_m$ (an empirical relation), where V is the rock's velocity, φ its porosity, V_f the velocity of the fluid in the rock's pores and V_m the velocity of the rock matrix. Gardner's relation is $\rho = kV^{1/4}$, relating density ρ to velocity through a constant k. (After Sheriff, 1984)

$$V_a = \frac{Z_1 + Z_2 + Z_3}{t_1 + t_2 + t_3} + \cdots \frac{Z_n}{t_n} = \frac{\sum_1^n Z_k}{\sum_1^n t_k}$$

(3) The *interval velocity* (V_i) is the estimated velocity of a layer between two reflectors,

$$V_i = \frac{Z_i}{t_i} \quad \text{or} \quad \frac{Z_2 - Z_1}{t_2 - t_1}$$

where Z_i and t_i refer to the thickness and one-way travel-time through a layer respectively. It may be determined from well velocity-logging, or more commonly from V_{rms} using the Dix equation (p. 174). Interval velocities are used to give information on the lithology and porosity of a layer, thus they are important in interpretation. They have 2–5% accuracy in areas of simple structure.

(4) The *stacking velocity* (V_{st}) is the velocity determined from the reflection move-outs and is used in stacking common-mid-point (CMP) records. The correct stacking velocity is that which removes NMOs most efficiently, converting a hyperbolic reflection into a line-up of arrivals and gives maximum summed amplitudes

after NMO correction (see p. 175). With the presence of significant dip, dip-move-out correction is necessary for the velocities.

(5) The *root-mean-square velocity* (V_{rms}) is defined by the equation

$$V_{rms}^2 = \frac{\sum_1^n V_i^2 t_i}{\sum_1^n t_i}$$

It could be obtained in a manual calculation from the slope of a t^2/x^2 curve at $x = 0$, i.e. $V_{rms} = V_a$ at $x = 0$. Generally, $V_{av} < V_{rms} < V_{st}$. V_{rms} was first defined by Dix (1955), who showed that where a series of constant velocity layers is used to model a structure, V_{rms} gives a better approximation to velocity in the NMO equation because of the straight ray paths in the model. Therefore using NMO values would give V_{rms} values in a layered model. V_{st} is an approximation to V_{rms} (rather than V_a) for flat parallel layers, or where there is a vertical gradient, but not in complex areas.

18.2 DETERMINING VELOCITIES

For determining average velocities and interval velocities, the most accurate method is in well velocity (sonic) logging and well shooting. In the latter (Figure 18.2) a number of surface shots or Vibroseis sources near the well are recorded at various depths in the well using a suspended hydrophone (constructed to withstand high pressures). Interval velocities are found by the differences between the reflection times from the bottom and top interfaces of the layer. For offshore wells, air-guns hung over the side of the platform are used as the sources. Recording a series of shots from one borehole to another hole (cross-borehole) is a recent innovation (Lines *et al.*, 1991).

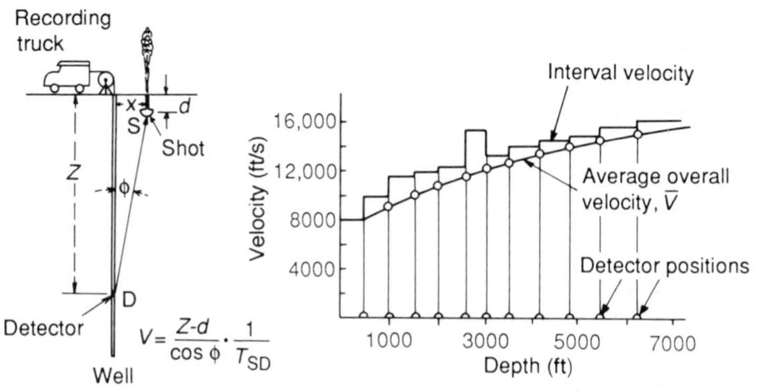

Figure 18.2 Well-shooting with typical average and interval velocity curves obtained. (After Dobrin and Savit, 1988)

Sonic velocity logging is the most common type of well velocity survey (Asquith and Gibson, 1982). The sonde lowered down the well includes one or two ultrasonic pulse transmitters and two or four receivers (Figure 17.2). Two transmitters, one at each end, about a third of a metre apart, enable tilt of the sonde to be eliminated and the effects of the well to be compensated. Sonic logs are used in mud-filled holes, and the sondes can also record surface shots to give average velocities, assuming flat layers, with the ultrasonic records giving interval velocities. They cannot be used where the well is cased — only below the casing.

Up-hole shooting uses deeper shot-holes (not wells), down to 70 m or so, to measure vertical velocities. This is carried out every few kilometres along traverses to determine the shallow velocities used in weathering corrections.

Velocity from surface shots: velocity determinations are done normally every 2–3 km or so along traverses, avoiding anomalous areas (Al-Chalabi, 1979; Robinson, 1983a). The various methods are described below; the first two, manual techniques, have now been replaced by the computer methods, but are still of interest.

The *t–Δt method* is based on the NMO equation and the move-out is calculated and removed at each distance.

$$\text{NMO}\,\Delta t \approx x^2/2V^2t_0 \qquad \text{or} \qquad V^2 \approx x^2/2t_0\,\Delta t$$

To remove the effect of dip, geophones need to be placed equally spaced on each side of the shot-point.

The *t^2/x^2 method* is the earliest procedure used. Since

$$t^2 = t_0^{\,2} + x^2/V_1^{\,2}$$

when the squares of the reflection times are plotted against the squares of the distances (Figure 18.3), a straight plot is obtained for constant velocity. Commonly, lower layers of different velocity produce a somewhat curved plot (higher terms). However, the t^2/x^2 method can give average velocities to a reflector to an accuracy of a few per cent with good-quality records and if static corrections are good. Dix (1955) showed that the t^2/x^2 plot at $x = 0$ (or short spreads) gives V_{rms}. This method and the $t - \Delta t$ method are the classic manual methods of velocity determination; they are now rarely used directly, but the latter is the basis of computer methods.

The Dix (1955) equation giving the interval velocity in layer $n\,(V_n)$ is

$$V_n^{\,2} = \frac{V_{\text{L}}^{\,2}\sum_1^n t_{\text{i}} - V_{\text{u}}^{\,2}\sum_1^{n-1} t_{\text{i}}}{t_n} = \frac{V_{\text{L}}^{\,2}t_{\text{L}} - V_{\text{u}}^{\,2}t_{\text{u}}}{t_{\text{L}} - t_{\text{u}}}$$

where V_{L} and V_{u} are the rms velocities, t_{L} and t_{u} times to the lower and upper layers (below and above the layer of interest respectively) and t_{i} the interval velocities in various layers.

The calculated V_{i} is not as accurate as V_{rms}; the smaller t_{i}, the less

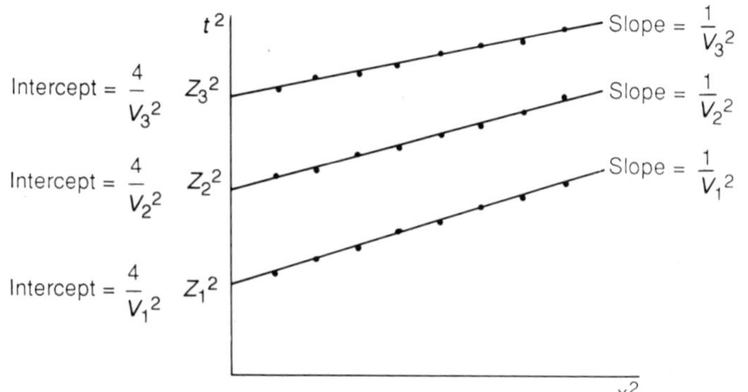

Figure 18.3 The t^2/x^2 plot for determining average velocities to reflectors manually. (After Dobrin and Savit, 1988)

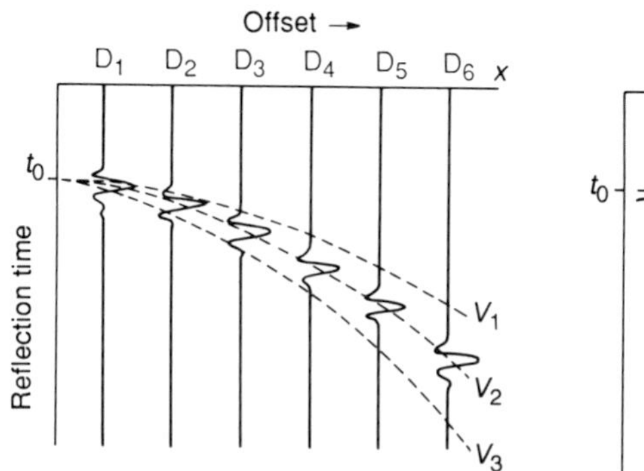

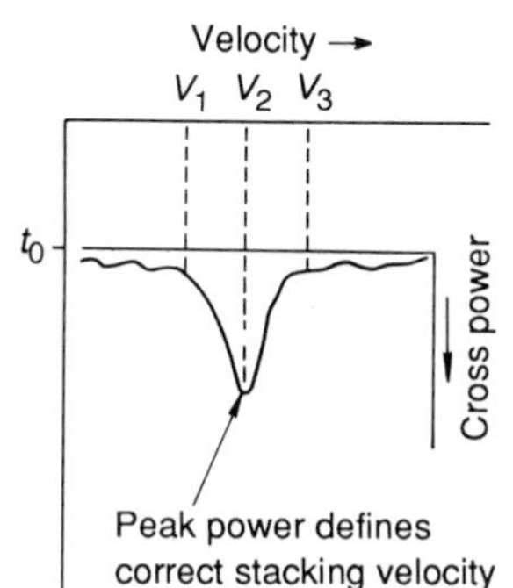

Figure 18.4 The velocity spectrum method of computer velocity determination for reflection data. Reflection data are corrected for move-out using a range of velocities. The stacking velocity chosen (V_2) is that which gives maximum power from the stacked events; it is the most efficient in removing move-out. (After Robinson and Coruh, 1988)

accurate, i.e. the thinner the layer, the less accuracy in V_i. Also note that strictly the Dix formula only applies to flat layers of constant thickness and velocity.

18.3 COMPUTER METHODS

These are the standard procedures for velocity determinations today, using the large amount of data available from CMP shooting. The methods use move-out values and thus the change in reflection time with distance (as we saw earlier, in the NMO correction), to determine the stacking velocity. The equation $t_x{}^2 = t_0{}^2 + x^2/V^2$ applies where t_x and t_0 are the time at distance x and zero respectively, and V is the stacking velocity. The stacking velocities, which are more closely related to V_{rms} than to average velocity (Cordier, 1985), may be converted to interval velocities using the Dix equation.

The most common computer technique uses the velocity spectrum (Taner and Koehler, 1969; Al-Chalabi, 1979; Yilmaz, 1987). The stacking velocity used in the NMO correction to CMP gathers is varied and the velocity giving the maximum sum for each CMP gather adopted (Figures 18.4 and 18.5). This is done at each time down the trace. Adding trace amplitudes for different CMP distances,

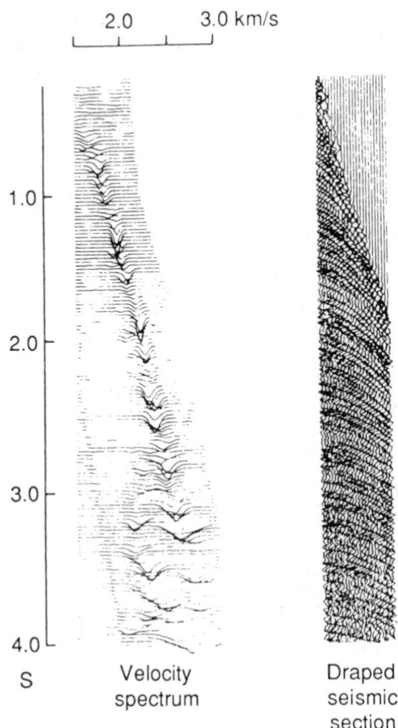

Figure 18.5 The velocity spectrum gives the stacking velocity versus reflection time down through the record. (After Taner and Koehler, 1969)

amplitudes will reinforce to give maxima at the correct velocities. For other velocities they will not add, but interfere. So velocity versus time (or depth) down the record can be chosen. A refinement of the method is the measuring of 'coherence' along hyperbolic trajectories (Sheriff and Geldart, 1983). Anomalously low velocities appear in the results, due to multiples (Figure 18.5) giving similar values of velocity as for primary reflections for greater times. We have assumed no dip; where dip is appreciable this has to be corrected for.

An alternative computer method is the *velocity scan* in which a series of fixed velocities is used to correct the move-outs and a separate record is made for each velocity (Figure 18.6). That is, the same velocity is used all the way down each record, for computing move-outs at various times. If the velocity is too high or low the reflection plot will be curved, usually quite strongly. The correct velocity for each time gives a horizontal reflection. So the velocity is easily plotted against time down the records.

To summarize computer velocity determination, obtain the stacking velocity by velocity spectra or velocity scan methods. This gives V_{rms} (approximately) from which the interval velocity V_i can be estimated by the Dix formula. A series of good V_i values enables estimation of the average velocities, plus depths and thicknesses, but V_i is not as accurate as the stacking velocities. Also there can be problems with dipping reflectors and horizontal velocity changes (see Dobrin and Savit, 1988, p. 252).

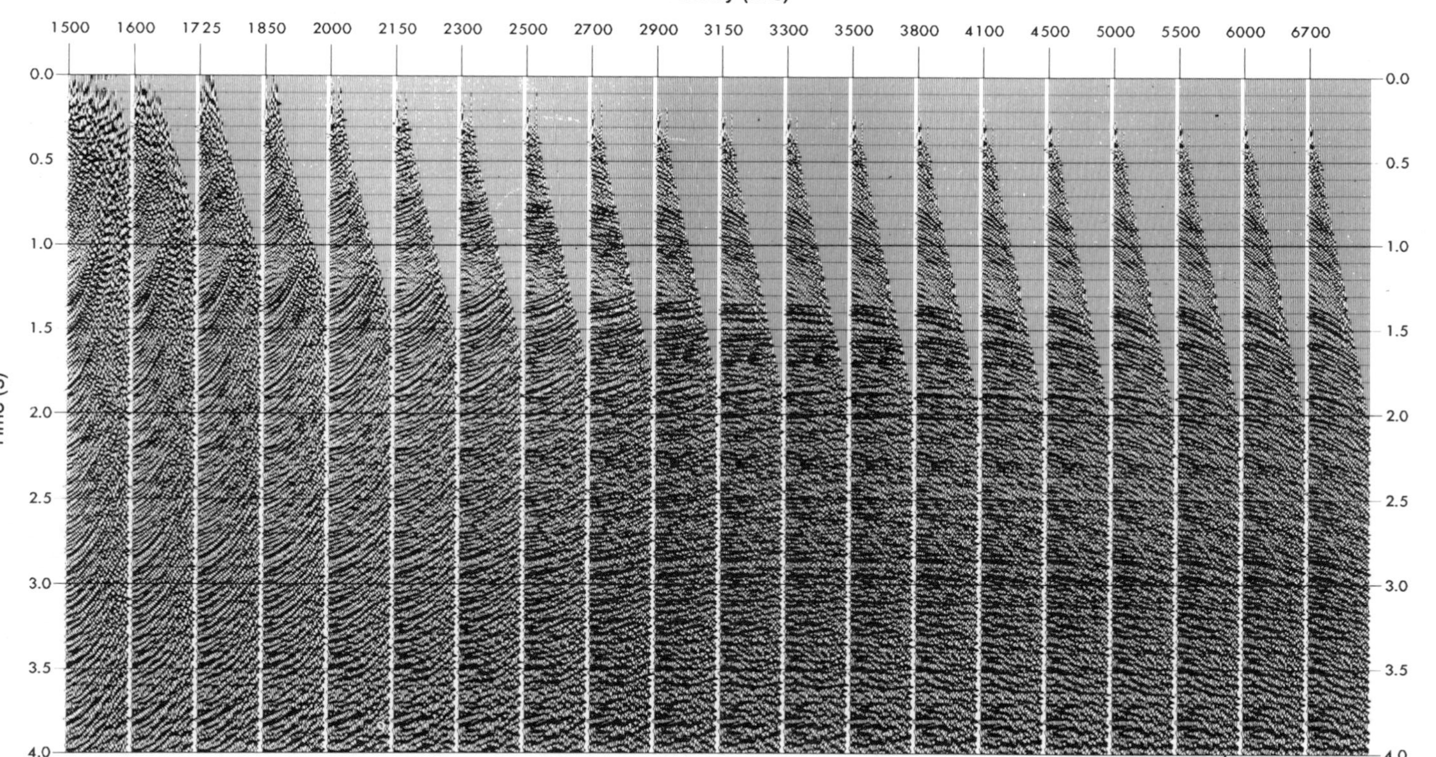

Figure 18.6 The velocity scan method of velocity determination. Each record strip from left to right is corrected for move-out at a successively higher velocity (fixed for each strip). The velocity appropriate for each time is where the reflectors are made horizontal (NMO well removed). (Courtesy of Richard Goto and Schlumberger Geco-Prakla)

Chapter 19
Multiple reflections

Multiple reflections are often one of the most serious sources of noise and may not be obvious to the interpreter. The simplest and most common is the surface multiple (e.g. Figure 16.1). It arrives at twice (or some other integral of) the travel-time of a normal primary reflection from the same reflector, unless there is a lot of dip, and so can be recognized by this double time. Also, multiples have larger move-outs than deeper reflections arriving at the same time, since their path is in lower-velocity upper layers compared to the deeper reflections. Thus a NMO stack is an efficient way of reducing deep multiples, and frequency filtering may also help as their frequencies tend to differ a little. Their dip move-outs are about twice that of the primary reflection with the same time. Other types of multiple are shown in Figure 19.1.

Multiple reflections occur where reflection coefficients, and so acoustic impedance (ρv) differences, are large, such as in weathered and aerated layers on land. Their amplitudes are proportional to the products ($r_1 r_2$) of the reflection coefficients of the interfaces, e.g. they may be prominent in marine surveys (Figure 19.2) where an ocean bottom such as limestone has a high-reflection coefficient (e.g. 0.4).

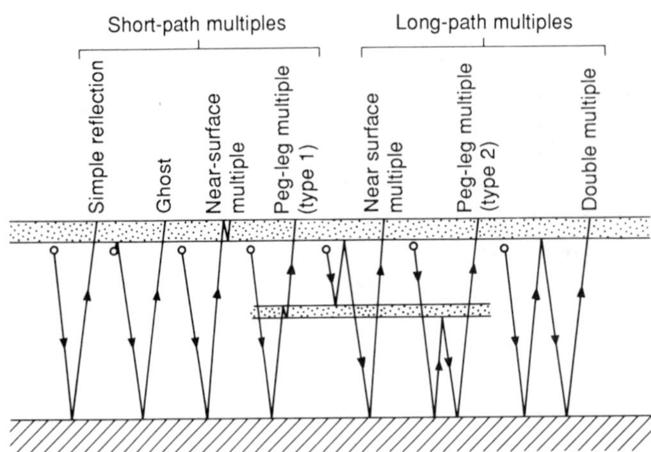

Figure 19.1 Various types of multiples. (After Sheriff, 1984)

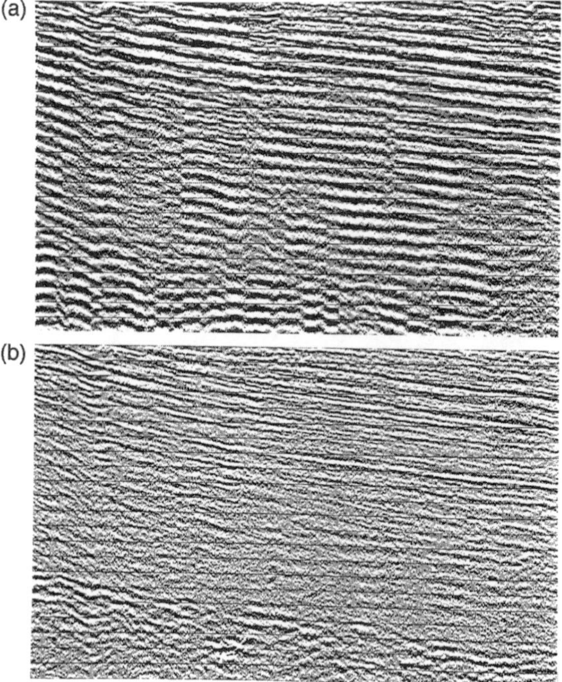

Figure 19.2 (a) A record dominated by multiple reverberations (ringing). (b) Same section after deconvolution. (Courtesy Schlumberger Geco-Prakla)

Ocean surfaces have a coefficient of -1 (causing a phase reversal), thus r_1r_2 in this case is -0.4. Some oceanic areas produce seismic records swamped by multiples, so-called 'ringing' records, caused by reflections bouncing back and forth (reverberating) in the water layer where the ocean floor has a high reflection coefficient, or on land reverberating in the weathered layer (Werth *et al.*, 1959).

A *long-path multiple* is one with a longer travel path than the primary reflection and so appears quite separately. *Short-path multiples* are produced by thin layers and mainly affect waveform through interference effects, lengthening the wavelet. 'Peg-leg' multiples are another common form (Figure 19.1).

Ghost reflections are reflections from the surface or the base of the weathering produced by a shot deeper than a quarter wavelength ($\lambda/4$) below it. The ghost (Figure 19.1) does not merge with the primary but produces a false reflection shortly following the primary. For shot depths less than $\lambda/4$ this can be beneficial, adding to the primary amplitude. Ghosts also affect waveform. Constant shot depths are used (9–15 m) so that the ghost adds to the primary.

There are three main methods of reducing multiples (see Yilmaz, 1987; Wardel and Whiting, 1989; Calvert, 1990; Hardy and Hobbs, 1991), but present techniques are not completely effective.

(1) CMP stacking (using the move-outs of the primary reflections).
(2) Velocity filters applied before stacking, e.g. in frequency/wave-

number (f–k) processing (see p. 169). This apparent velocity filtering attempts to distinguish between near-vertical reflections, for which the apparent velocities are very high, even for quite steep dips, and surface waves and multiples whose rays are not near-vertical.

(3) Deconvolution methods: these have been used to reduce short-period multiples which dominate records (Arya and Aggarwal, 1962; Silvia and Robinson, 1979; Robinson and Treitel, 1980) by cancelling the filter effects of the water layer (inverse filtering).

Chapter 20
Seismic migration

20.1 INTRODUCTION

Migration is an important method of improving the seismic image of the sub-surface, plotting dipping reflectors much more accurately and focussing diffractions. An unmigrated stacked section shows reflectors plotted under the mid-point; however, with dip the reflector would actually be up-dip and dips actually greater than plotted in an unmigrated section (Figure 20.1). The true dip θ_t is given by $\sin \theta_t = \tan \theta_a$, where θ_a is the apparent dip. Thus migration attempts to position reflectors at their true positions. Reflection points are moved (migrated) both vertically and horizontally and always up-dip. Good migration, being a focussing process, also collapses diffraction hyperbolae to their points of origin (Figure 2.18) (Hood, 1979; Grau and Lailly, 1989). A number of pioneering papers are published in Gardiner (1985).

The distortion in an unmigrated section increases with dip, and for

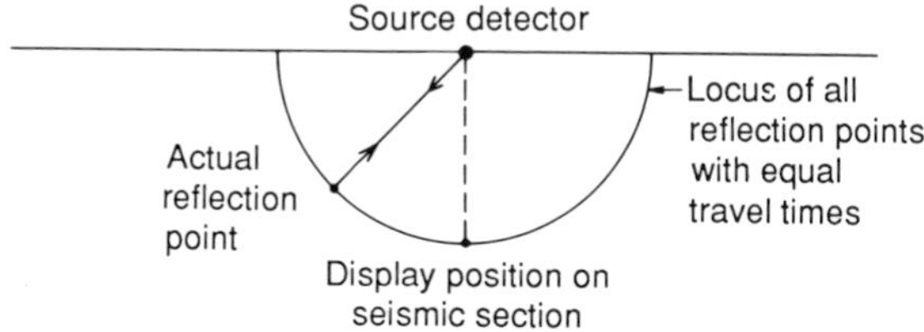

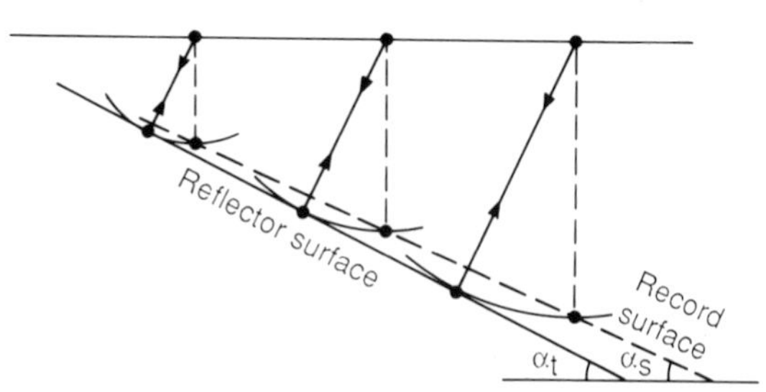

Figure 20.1 The principle of migration. (a) For a given reflection time, the reflector may be anywhere on a circle. On a non-migrated section this is mapped below the common mid-point. (b) A plane reflector and how it would be plotted without migration. (After Kearey and Brooks, 1991)

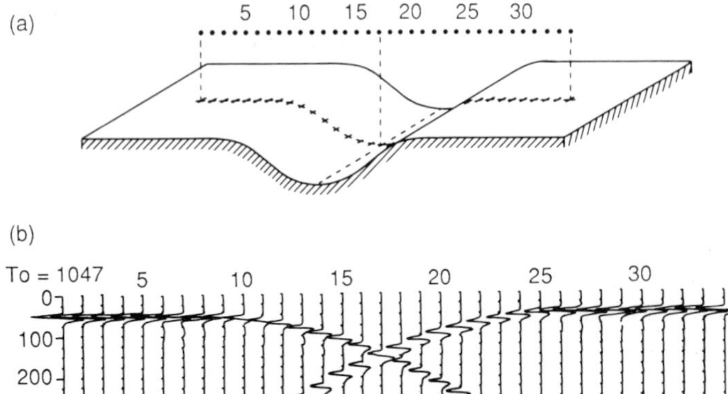

Figure 20.2 (a) A model of a simple syncline, and (b) an unmigrated reflection record showing the bow-tie and triple-arrival effect. (After Am. Assoc. Pet. Geologists)

dips greater than 5° the distortion makes an unmigrated section hard to interpret accurately. Anticlines appear broader and less curved than they are; synclines less broad (Figure 2.19). If the curvature in a syncline is greater than that of the wave-front, three reflections will occur instead of one (the bow-tie effect), a 'defocussing' (Figure 20.2). Thus the focussing properties of migration are desirable, producing a more perfect imaging of the sub-surface.

While deconvolution sharpens wavelets relative to time, migration sharpens or collapses a wavelet with respect to the horizontal distance. The migration process is an inverse scattering problem, seen in other branches of physics (see Menke, 1984). It used to be expensive but is now a routine process and most necessary in target areas where structures are steeper and more complex. It is important in locating the best point to drill.

20.2 MANUAL MIGRATION

Computer techniques are much more efficient and sophisticated and are now used routinely (see below), but the old manual methods are instructive. If the data are NMO-corrected the reflection point would be on a circle (for constant velocity) centred at the SP-geophone mid-point (Figure 20.1). Thus a series of such circles would map the reflector as the envelope of the arcs. Originally all migration methods were done manually, using ray tracing, diffraction and wave-front diagrams and plotting arms, based on reflection times and move-out times and known or postulated velocity structures (Hagedoorn, 1954). Wave-front charts (Figure 20.3) can be used for manual migration where NMO corrections have been applied (see Dobrin

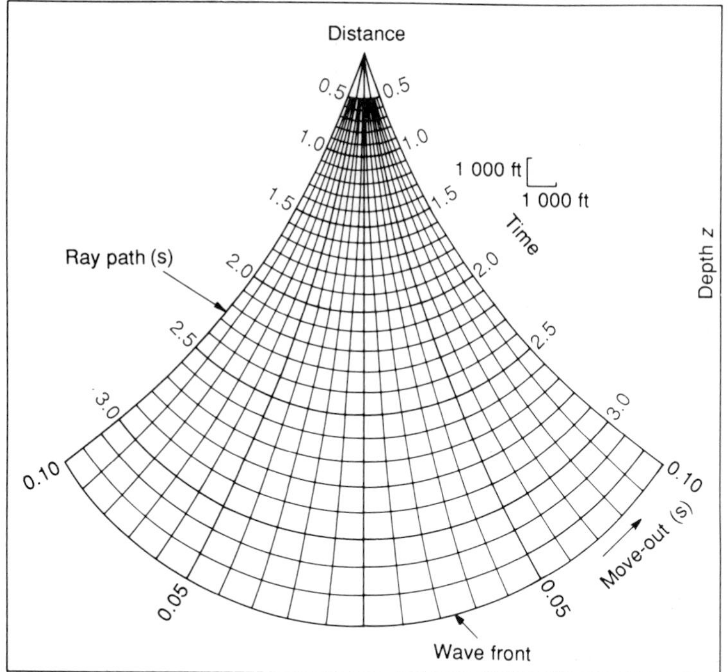

Figure 20.3 A wave-front chart for a linear increase of velocity with depth. Travel-time and move-outs are shown. (After Dobrin, 1976)

and Savit, 1988, p. 268). The travel-time and NMO specify the ray, and the reflector and its dip (normal to the ray) can then be plotted. This was the commonest manual method although it tends to over-migrate.

An important point is that this method, and other two-dimensional methods, assume that we are traversing perpendicularly to the maximum (true) dip direction. Otherwise, three-dimensional effects can occur, such as reflections and diffractions from the side ('side-swipe'). Low-noise, good statics corrections and knowledge of velocities are also required, together with adequate sampling and small station spacing (Stolt and Benson, 1986).

20.3 COMPUTER MIGRATION

Computer migration is standard practice now, if the cost warrants it (Robinson, 1983b; Sheriff and Geldart, 1983). A clear example of computer migration is the old wave-front migration method. This was the first computer method, and is not used today. It is the computerized method of drawing the envelope of circles. In effect, the computer is programmed to draw the circles centred on each shot-point with radii equal to the appropriate travel-times. Each cell of the section is weighted according to the amplitude and number of circles passing through the cell. The wave-front method works if

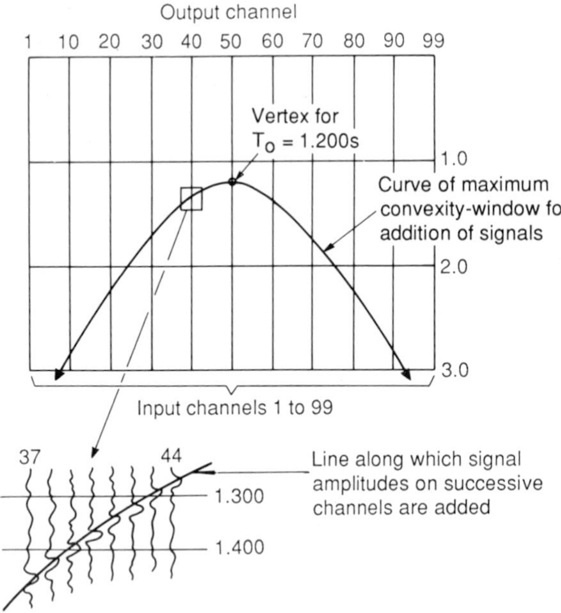

Figure 20.4 The principle of computer migration using the diffraction curve (curve of maximum convexity). Signals are added along the curve at each time it intercepts a trace. Such additions are made for a series of such curves with an apex at each sample (record) time. Where energy has been diffracted along such a curve, significant amplitude will be plotted at the apex. (After Dobrin, 1976)

velocities are well known, but tends to smear the records and it was only used where the signal-to-noise ratio was high.

Diffraction stacking is an interesting computer migration method employing diffraction curves; again, it is no longer used. A reflector can be thought of as being made up of a multitude of diffracting points or scatterers. A diffraction curve connects positions on the records where energy arrives that would be diffracted from a point on a reflecting surface (Figure 20.4). With known or assumed velocities a diffraction curve can be calculated for each recorded point on a trace. The energy from the diffracting point can thus be found by summing recorded amplitudes along the calculated diffraction curve and placing the sum at the apex. Where diffraction has not occurred, of course, data tend to cancel out and sums are near-zero. If the section is searched along all possible diffraction curves in this way, diffractions can be collapsed and reflections can be migrated, as a reflection is the sum of (or interference between) diffracted waves from all the points on the reflector.

For a 3 s record, for example, summations would be made along a diffraction curve at each sample time (e.g. each 2 ms), i.e. along 1500 'hyperbolic' curves for each channel. There may be 100 channels requiring 150 000 curves to be calculated — hence the computer cost.

The diffraction curve is a hyperbola where the velocity is constant. If the velocity varies with depth, a similar curve has to be calculated. However, in such diffraction stacking the diffracted events tend to not be in complete phase, so some energy is lost. This is overcome in the more complete wave equation methods.

Wave equation migration computer techniques are the most common

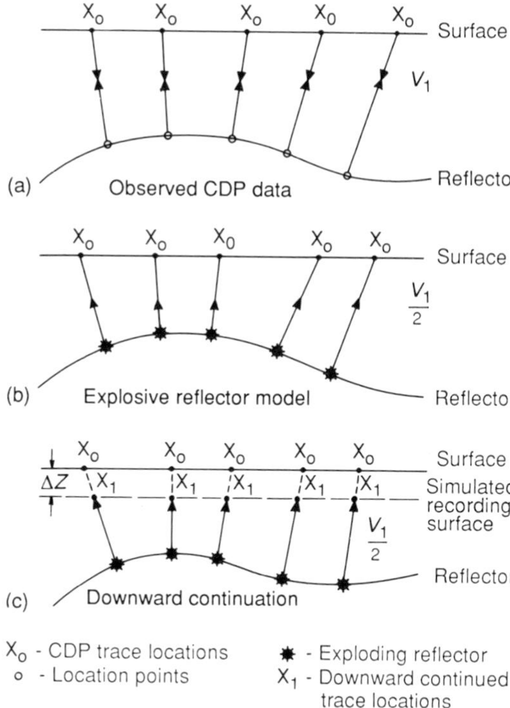

Figure 20.5 (a) The CDP stacked section simulates data corresponding to two-way transmission along paths normal to a reflecting surface. (b) The exploding reflector model for wave-equation migration. (c) Downward continuation of receivers to depth ΔZ. (After Johnson and French, 1982)

methods in use today and are analogous to downward continuation in gravity and magnetic computations (Stolt and Benson, 1986). The different techniques are summarized in Dobrin and Savit (1988), but also see Kleyn (1983), Claerbout (1985) and Berkhout (1987). Some of the different kinds are those involving calculations in the time–distance domain (time migration), in depth and distance (depth migration), or the frequency–wave number (*f*–*k*) domain.

Time migration begins with the recorded CMP stacked amplitudes and phases, and continues the wave-field downward. At each depth the wave is calculated at $t/2$, backward in time; if there is significant energy there it must be on a reflector. Alternatively, the method visualizes sources on the reflector transmitting upward and uses the wave equation to calculate the distance moved and the phase changes (Figure 20.5). In the finite-difference method (Claerbout, 1985) an approximation to the wave equation is used.

One time-domain migration method is *Kirchoff migration* (Schneider, 1978). The integral solution of the wave equation is used to sum over the diffraction hyperbola, as in the diffraction stacking method described above, but also corrects for phase shift. This requires a reasonably good model of the velocities with depth (within 10%) and only small velocity changes horizontally. The 'migration velocities' used come from rms velocities and other data and are corrected for dip. It handles dips to 90°. With steep dips, accuracy of migration requires good migration velocities.

Frequency domain (*f–k*) migration (Chun and Jacewitz, 1981; Stolt and Benson, 1986) is popular as it is fast and cheap, transforming data into frequency (f) and wave number (k, the inverse of wavelength). The method migrates by convolution. The data are first 'stretched' to approximate a constant velocity, then transformed into the *f–k* domain, migrated and transformed back for a time–distance plot. It handles large changes in velocity but handles dips to 80° only.

Depth migration, in which times have been converted to depths, should be used when there are strong horizontal velocity changes, producing more accurate imaging than time migration, but it requires a good velocity model and is costly (Dobrin and Savit, 1988). The output is in distance versus depth.

Seismic modelling: if velocity variations cannot be represented by a function or average velocities, because of very steep dips and/or velocity jumps, models of the sub-surface may be set up as a first approximation using the unmigrated data. Ray tracing is then carried out and models compared iteratively with the measured times (Kennett and Harding, 1985). Seismic modelling is also used to correlate with well-log data and to check interpretations. Velocity anisotropy can be included in models. In some rocks, particularly shales, horizontal velocities may vary from those measured vertically by 50%.

Three-dimensional migration: migration in 3D is ideally necessary, particularly with complex structures, being more accurate and removing the effects of side reflections. This method is commonly carried out over known or suspected hydrocarbon deposits (e.g. Dahm and Graebner, 1982). The cost of 3D surveys and the processing is of course much greater. Detailed orthogonal traverses are necessary. Three-dimensional migration could be carried out by summing amplitudes over a hyperboloid at each time but in practice a succession of 2D migrations is commonly made in two perpendicular directions (Gibson *et al.*, 1983; Claerbout, 1985). Increased computer power has made 3D migration more possible.

Migration after stacking (poststack) is usually satisfactory with fairly strong lateral velocity variation, but sometimes not and *prestack partial migration* is then necessary (Yilmaz and Claerbout, 1980). Over complex structures with strong lateral velocity variations and dips it may be essential to use prestack depth migration and an accurate velocity model. Prestack migration takes much more computer time and much more expense, but has the possibility of revealing incorrect velocities and also of retaining amplitude variation with offset (p. 191), used to recognize lithology.

Chapter 21
Presentation and interpretation of data

21.1 PRESENTATION

Seismic 'sections' are commonly printed as black and white 'wiggles', variable density of trace, variable area or a combination of wiggle plus one of the other two (e.g. Figure 15.10). Sections are usually plotted against time, particularly for the early interpretations. For sections plotted versus depth, reasonable velocity values are necessary and are preferably done after migration. The deeper portions are stretched by the increasing average velocity with depth and appear with longer wavelengths. This accentuates the Earth low-pass filter effect (the greater attenuation of higher frequencies) but is useful in illustrating the reduction in resolution with depth. In addition, coloured sections are becoming quite common and are now being used to show clearly seismic characteristics such as amplitude, dip and reflection correlation (Marillier *et al.*, 1990).

Figure 21.1 Special computer work-stations are now used for interactive processing and interpretation of data. (Western Geophysical)

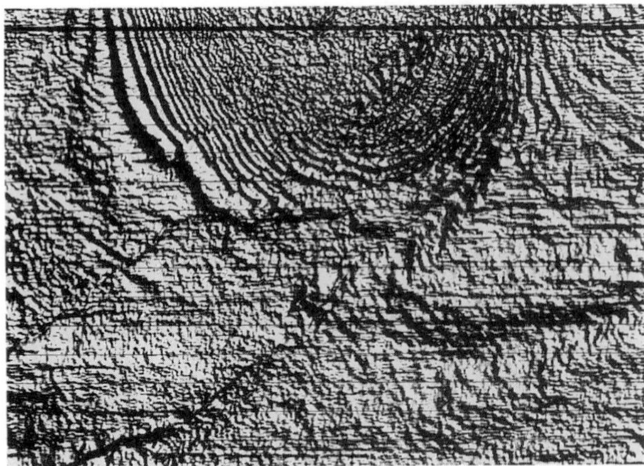

Figure 21.2 A 'time slice' from 3D data from the Gulf of Mexico. The slice cuts the data at 1 s. Note the evidence for the salt dome and faulting. (Courtesy of Western Geophysical)

Special computer work-stations are now used for interactive processing and interpreting of data (Figure 21.1), e.g. for manual or computer picking of reflections, contouring, migration and modelling (Coffeen, 1990). This is particularly useful for 3D surveys. The data are put into the computer from the magnetic tapes. The use of work-stations also greatly aids the presentation of sections, their filing, the storage of picked times, their plotting and conversion to depths. Processing and interpretation can now even be done in the field.

With detailed three-dimensional data, results can be presented as a series of slices down through the area (Figure 21.2), so-called time slices, showing the changing structures with depth and time. Any required vertical slice through the data is also possible with 3D data and computer manipulation and display (Figure 1.11). Muting is the deletion of shallow (very early) reflection data which has been distorted in NMO correction and/or by refraction and surface waves.

21.2 GEOLOGICAL INTERPRETATION OF REFLECTION DATA

One of the most important aspects of geophysics, and sometimes the most difficult, is the geological interpretation of the physical data (McQuillin *et al.*, 1979; Coffeen, 1984; Badley, 1985). In oil and gas exploration one is looking for a structural or stratigraphic trap which may contain hydrocarbons in a porous stratum (Figure 21.3). The structural traps are much more easy to recognize in seismic data than are stratigraphic traps, e.g. anticlines and faults are simpler to identify than pinch-outs and unconformities.

A seismic section is really an interference pattern produced by the addition of reflections from the numerous interfaces in a sedimentary column, many of which are thinner than the seismic wavelengths. Also phase shifts are introduced by the Earth, the instruments and by

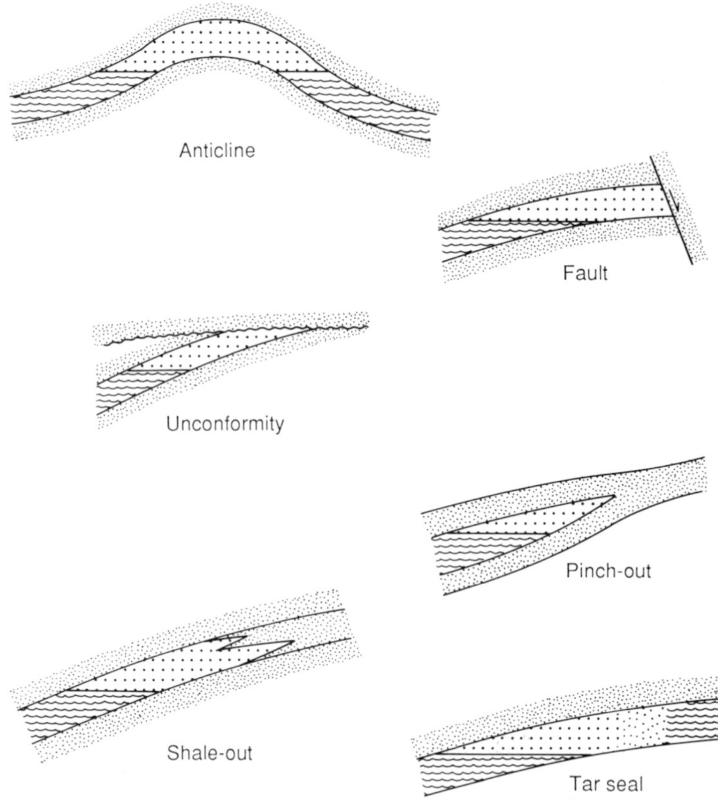

Figure 21.3 A schematic diagram of some types of hydrocarbon traps. (After Dikkers, 1985).

digital processing of data, usually as phase lags. For example, geophones are velocity recorders, not amplitude recorders, thus introducing phase shifts.

'Reflections' occur which are not produced by a change in lithology. Horizontal or vertical velocity variations cause distortions in time sections and use of an incorrect stacking velocity causes errors in structural interpretations. Hence the need for correlation of reflection data with velocity (sonic) logs down wells to check interpretations, depths, etc. The first step after constructing the seismic sections is to identify the reflectors geologically by relating them to well data, particularly sonic logs (Figure 21.4), and to surface geology if appropriate. The analysis of seismic data recorded from one borehole to another (cross-borehole) has also recently been used giving good resolution (Lines *et al.*, 1991).

Synthetic seismograms are useful for comparing with the seismic sections and so aid in interpretation, i.e. to see which reflections are significant and should be mapped, and which are multiples. The synthetic seismograms are obtained by convolution of the source waveform with reflectors modelled from well log data or vertical seismic profiling (Kennett *et al.*, 1980; Cassell, 1984; Balch and Lee, 1984; Hardage, 1985) (Figures 21.5 and 21.6).

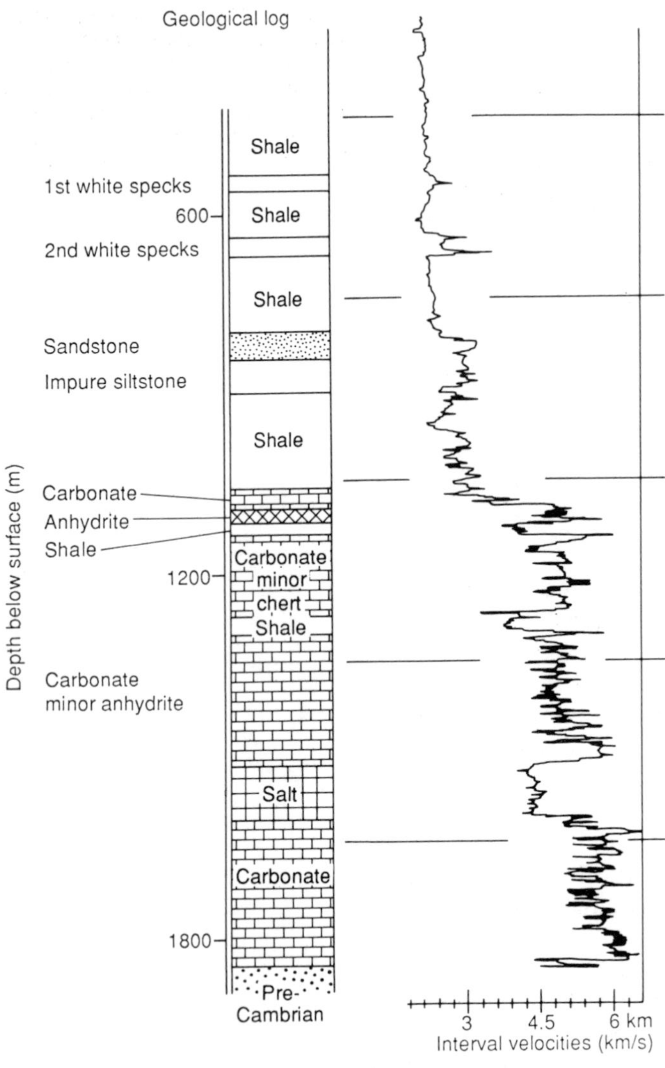

Figure 21.4 A velocity log with the geological section. Note frequent changes in velocity in the same formation. (After Grant and West, 1965)

Relative depths are fairly accurate, but for absolute depths well-ties are necessary. Note that resolution falls off with depth, e.g. it is about 10 m at shallow depths and 70 m towards maximum depths because of higher velocities, low frequencies and long wavelengths. If a layer thins or 'pinches out' to about three-eighths of a wavelength, the two reflections from the top and bottom merge into one and the layer is no longer resolved (Figure 15.5). On the other hand, high-amplitude reflections may come from layers $\frac{1}{4}\lambda$ thick or less due to addition of reflections.

Horizontal resolution is less than vertical and also decreases with depth. It is governed by the fact that reflection energy comes from Fresnel zones of radius $V/4\,(t/f)^{1/2}$, where V is the average velocity, t is the two-way time, and f is frequency (Figure 21.7). Outside the Fresnel

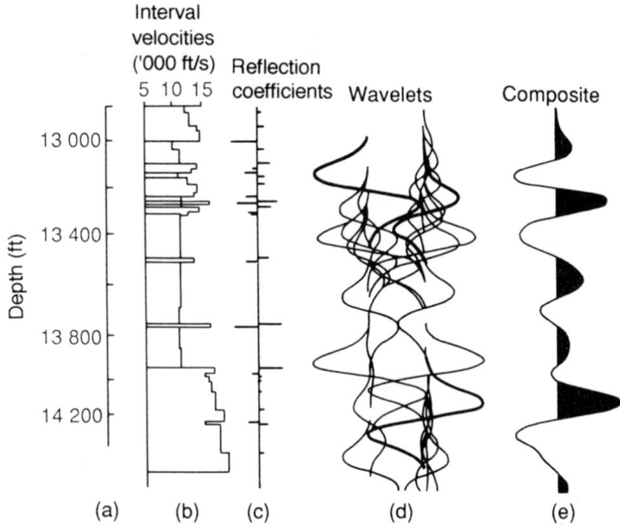

Figure 21.5 The construction of a synthetic seismogram: (a) depth values, (b) interval velocities, (c) reflection coefficients calculated, ignoring densities (a common practice), (d) individual wavelets placed at each reflector time with the appropriate amplitude, and (e) the sum of the wavelets. (After Vail *et al.*, 1977)

radius interference reduces amplitudes, and structures smaller than the resolution would not be picked (Figure 21.8). As seismic waves have long wavelengths (10^9 times the length of light waves) reflection occurs over a large Fresnel zone. For reflector dimensions rather smaller than the Fresnel zone the response is that of a diffracting point.

Interval velocities can be used to help identify lithology (Table 2.1) (Domenico, 1984) and also densities of the layers (Gardner *et al.*, 1974). However, interval velocities are not accurate if layers are not horizontal. The porosity of the rock also affects velocity, which can be useful (Figure 21.9). Extra information on lithology and porosity might be gained from S-wave velocities as the V_p/V_s ratio has been claimed to be a useful discriminator (Domenico, 1984; Dohr, 1985; Danbom and Domenico, 1986; Johnston and Christensen, 1992). However, Ikwuakor (1988) has discussed some complications. S-wave anisotropy can be related to fractures.

Another technique used to identify lithology and indicate hydrocarbons is the interpretation of the variation of reflection amplitudes with angle of incidence or offset (AVO) using pre-stack data (Ostrander, 1984; Ursin and Dahl, 1992). Stacking destroys amplitude information. Amplitudes are related to the elastic constants and pore fluids and can be interpreted in terms of lithology and porosity, e.g. gas in sediments may show an increase of corrected amplitude with offset whereas basalt shows a decrease. An interesting earlier development was the occasional direct detection of gaseous hydrocarbons by the strong reflections (bright spots) and horizontal reflections (flat spots, gas over water or oil) from low-velocity gas boundaries (Figure 21.10) (e.g. Backus and Chen, 1975). Other lithologies can also produce bright spots, e.g. salt, carbonates and volcanics.

Where a reflector dies out along a traverse, a 'phantom' horizon is

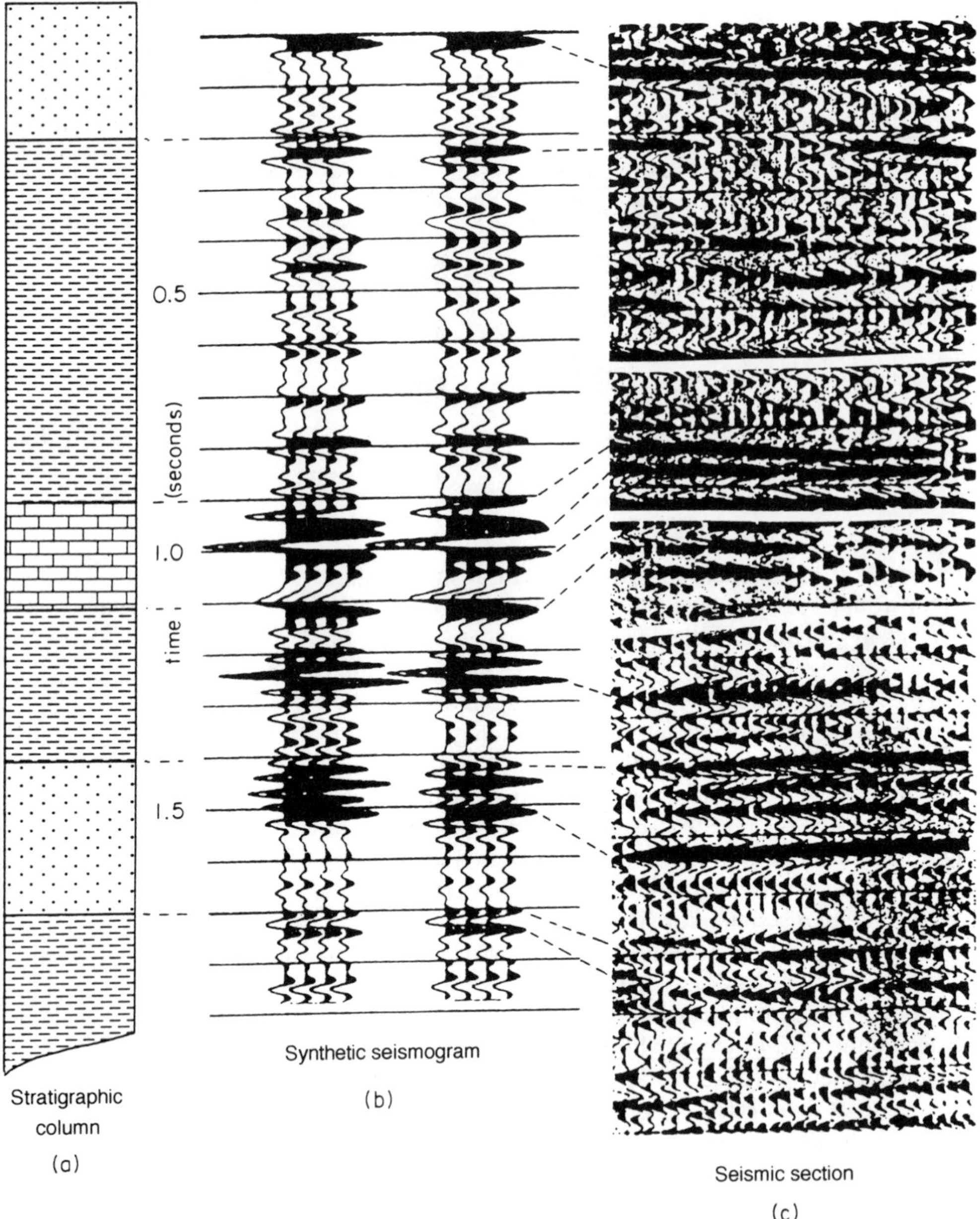

Figure 21.6 An example of a synthetic seismogram: (b) with the stratigraphic column (a), and the actual seismogram (c) for comparison. (Courtesy of Edwin S. Robinson)

plotted parallel to other nearby reflections to join up with visible reflectors further on. A phantom horizon can be checked by tracking around loops of traverses to make sure you are still on the same reflector (closure tests). An error of one cycle in picking a reflection may change the calculated depth by 30–60 m, thus all picked reflectors should be checked around a loop. Interpretation can be

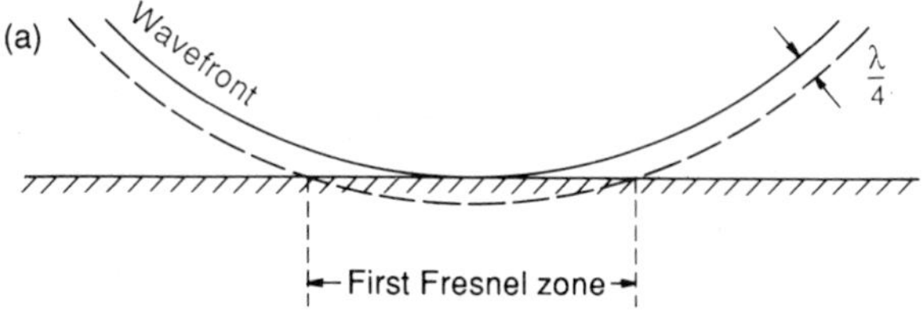

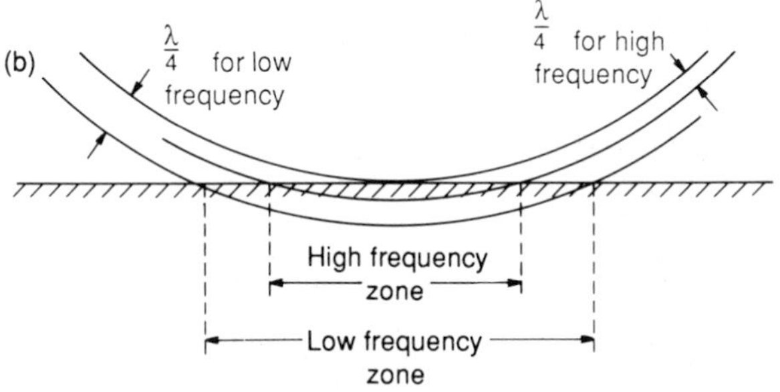

Figure 21.7 The Fresnel zone which defines the horizontal resolution for a certain frequency or wavelength. The area of the reflector from which energy can return within the next half-cycle is given by the circle that the wave-front makes on the reflector $\frac{1}{4}$ wavelength later. The Fresnel zone is larger for longer wavelengths (lower frequencies). (After Sheriff, 1980)

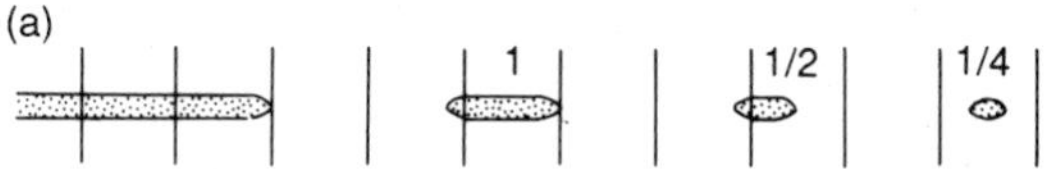

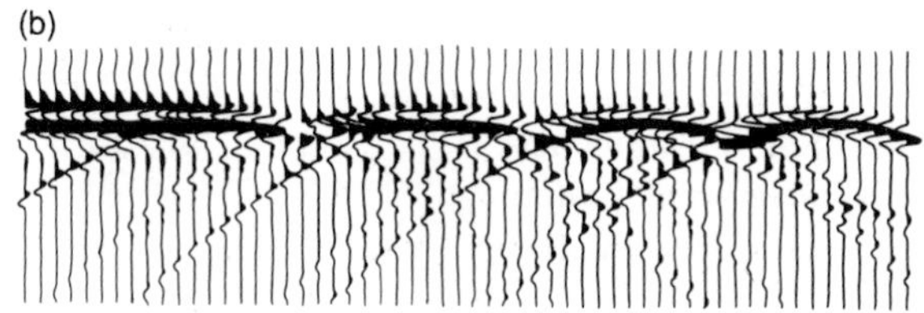

Figure 21.8 Reflection from reflectors of various small size: (a) cross-section of the model; vertical lines are spaced by the Fresnel zone size; (b) seismic section from such a model showing the diffraction hyperbolae. (After Neidell and Poggiagliolmi, 1977)

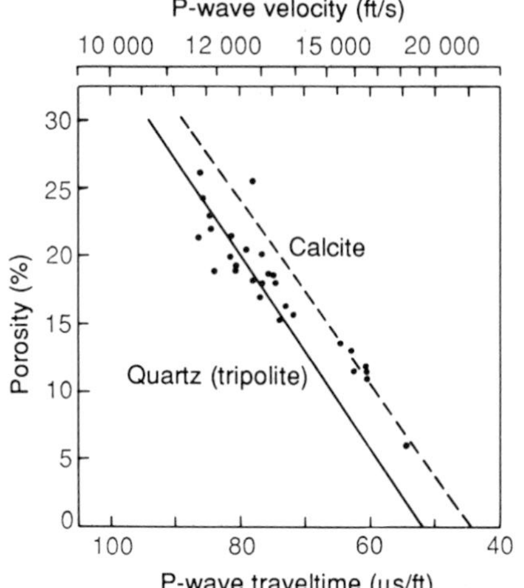

Figure 21.9 Velocity versus porosity. The lines are the time-average equations for quartz and calcite. (After Gardner *et al.*, 1974)

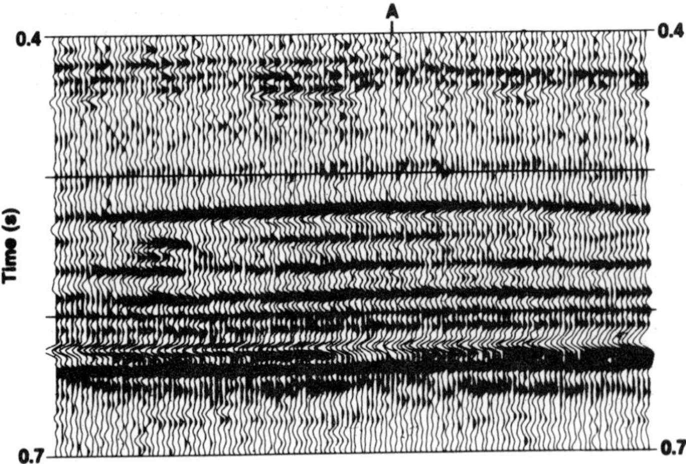

Figure 21.10 Strong ('bright spot') reflection at 0.52 s. The well at A produces gas from sand at that level. (Courtesy of Western Geophysical)

assisted by horizontal compression of the record (a vertical exaggeration of perhaps 5–20 times) by computer, particularly for regional data. This helps in the recognition of patterns and thickness variations.

Depths are plotted on a base map, also seismic lines, shot-points, wells, shorelines, roads, gravity anomalies, known faults, etc. Then isopach maps, showing the estimated thicknesses of each layer, can be constructed from the depth differences.

The most common structures mapped for oil are anticlines and faults (Figure 21.3). Diffraction effects can produce apparent anticlines if the data have not been migrated, but this may be recognized. Faults can also be recognized fairly easily in some cases, e.g. by the diffractions they produce at the edges of layers.

Salt domes may show up clearly (Figure 21.11). These are thick salt deposits buried under unconsolidated sediments which become buoyant as sediments above become denser with compaction. The salt then acts like a very slow fluid, doming up and possibly piercing sediments. Oil may be trapped in a narrow belt about the dome. Igneous plugs can have a similar appearance to salt domes. Gravity and magnetic data can resolve this as such plugs are more dense and usually more magnetic than sedimentary rocks.

Basement, the top of the hard igneous or volcanic rock underlying sediments, is usually distinct and can be recognized by a large number of diffractions (Figure 2.17).

21.3 SEISMIC STRATIGRAPHY

Seismic data have largely been used to investigate geological structures, but with the growth of greater detail and resolution in surveys, plus improved data processing methods, seismic stratigraphic methods

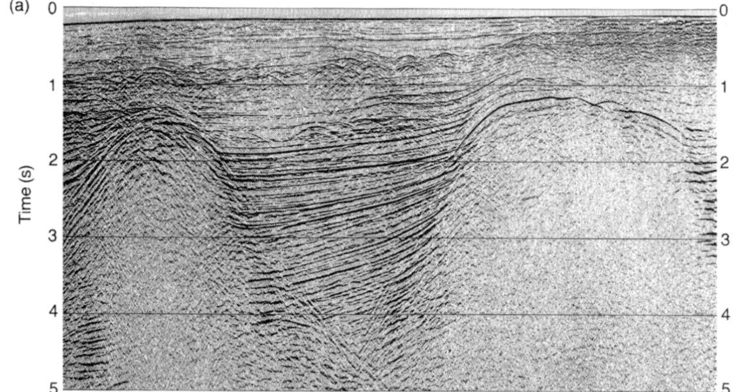

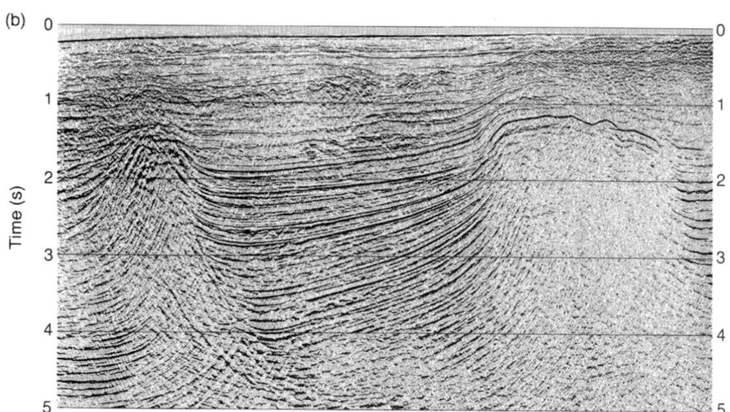

Figure 21.11 Salt domes on a section from the Gulf of Mexico. (a) An unmigrated stack; (b) a migrated stack. Note the diffraction hyperbolae from irregular surfaces below the very top layers in (a). (Courtesy of Soraya Brombacher and the Western Geophysical Co.)

have also been developed. Seismic stratigraphy involves the division of records into *seismic sequences* of related sediments, which can reveal the geological history of the area (Vail *et al.*, 1977; Brown and Fisher, 1980; Sheriff, 1980; Helbig and Treitel, 1987; Cross and Lessenger, 1988).

Reflections are interpreted to represent time (isochronous) surfaces (e.g. bedding planes) or erosion surfaces (unconformities). Related sedimentary sequences normally comprise strata bounded by unconformities produced by erosion and deposition in periods of uplift or sea-level change. The sequences are then interpreted for evidence of the environment of the sedimentary deposition. The importance of seismic data may often be as valuable in elucidating the environment and sedimentary history as in the direct location of hydrocarbons, e.g. changes in sea level, whether the area was a quiet (low-energy) sea, an energetic shore, a delta, a lake or a region of river channels. The application of seismic stratigraphy in basins around the world is claimed to have shown the distribution of major global unconformities and to allow recognition of worldwide patterns of sea-level change. Correlation with well log data and high resolution biostratigraphy has produced 'sequence stratigraphy'.

Appendix 1
Suggestions for further reading

Bolt (1993) and Eiby (1980) have produced good introductory texts on earthquakes, and Howell (1990) one on the history and development of general seismology. The US Geological Survey publishes a journal, *Earthquakes and Volcanoes*, aimed mainly at the general reader. Amongst the more prominent textbooks on earthquakes and general seismology are Richter's *Elementary Seismology* (1958), still a valuable descriptive text; Bullen's (1947) *Introduction to Seismology*, a largely theoretical treatment, the most recent edition being updated by Bolt (Bullen and Bolt, 1985); *Elastic Waves in Layered Media* by Ewing *et al.* (1957), a mathematical discussion. Other mathematical texts include Aki and Richard's (1980) *Quantitative Seismology, Theory and Methods* (two volumes), and Ben-Menahem and Singh's (1981) *Seismic Waves and Sources*, both advanced theoretical treatments; also Kennett (1983), *Seismic Wave Propagation in Stratified Media*; Pilant (1979), *Elastic Waves in the Earth*; White (1983), *Underground Sound, Application of Seismic Waves*; James (1989), *The Encyclopedia of Solid Earth Geophysics*; and Gubbins (1990), *Seismology and Plate Tectonics*.

All these publications deal with general seismology but not exploration seismology. This large subject is dealt with in the following: Sheriff and Geldart (1982, 1983), *Exploration Seismology* (two volumes); Waters (1987), *Reflection Seismology*; Coffeen (1986), *Seismic Exploration Fundamentals*; Sengbush (1983), *Seismic Exploration Methods*; Claerbout (1985), *Imaging the Earth's Interior*; Dix (1981), *Seismic Prospecting for Oil*; Berkhout (1987), *Applied Seismic Wave Theory*; Clay (1990), *Elementary Exploration Seismology*; Gadallah (1994), *Reservoir Seismology*; Yilmaz (1987), *Seismic Data Processing*; and the large number of texts introducing all geophysical methods, but particularly Dobrin's *Introduction to Geophysical Prospecting*, recently updated by Savit (Dobrin and Savit, 1988); Telford *et al.* (1990), *Applied Geophysics*; Kearey and Brooks (1991), *An Introduction to Geophysical Prospecting*; Robinson and Coruh (1988), *Basic Exploration Geophysics* and Parasnis (1986), *Principles of Applied Geophysics*. Sharma (1986), *Geophysical Methods in Geology*, in a good introductory text to both general and exploration geophysics and Sheriff (1984) has produced an *Encyclopedic Dictionary of Exploration Geophysics*.

Appendix 2
The Zoeppritz equations

To determine the relative amplitude of reflected and transmitted waves for any angle of incidence (Figure 2.15), the Zoeppritz equations must be used. Displacements and the stresses are continuous across the interfaces (Figure A2.1).

$$A_1 \cos e_1 - B_1 \sin f_1 + A_2 \cos e_2 - B_2 \sin f_2 = A_0 \cos i$$
$$A_1 \sin e_1 + B_1 \cos f_1 - A_2 \sin e_2 - B_2 \cos f_2 = -A_0 \sin i$$
$$A_1 Z_1 \cos 2f_1 - B_1 W_1 \sin 2f_1 - A_2 Z_2 \cos 2f_2 + B_2 W_2 \sin 2f_2 = -A_0 Z_1 \cos 2f_1$$
$$A_1 g_1 W_1 \sin 2e_1 + B_1 W_1 \cos 2f_1 + A_2 g_2 W_2 \sin 2e_2 + B_2 W_2 \cos 2f_2 = A_0 g_1 W_1 \sin 2i$$

where

$$Z_1 = \rho_1 \alpha_1,\ Z_2 = \rho_2 \alpha_2,$$
$$W_1 = \rho_1 \beta_1,\ W_2 = \rho_2 \beta_2,\ g_1 = \beta_1/\alpha_1 \text{ and } g_2 = \beta_2/\alpha_2$$

α and β are P and S velocities and i, e and f the incident, reflected and refracted angles for P and SV (Figure A2.1). Subscripts 1 and 2 refer to the top and bottom layers.

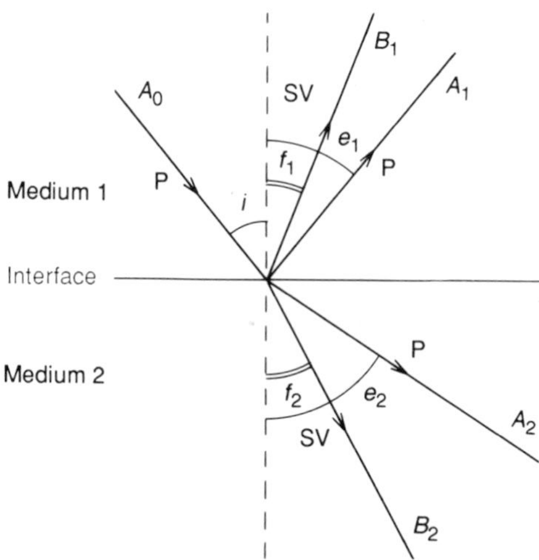

Figure A2.1 The symbols in the Zoeppritz equations for the amplitudes of reflected and transmitted (refracted) waves

References

Agnew, D. C., 1986. Strainmeters and tiltmeters. *Rev. Geophys.*, **24**, 579–624.

Agnew, D. C., 1989. Seismology: history. In James, D. E. (ed.), *The Encyclopedia of Solid Earth Geophysics*, Van Nostrand Reinhold, New York, pp. 1198–1201.

Aki, K., 1984. Asperities, barriers, characteristic earthquakes and strong motion prediction. *J. Geophys. Res.*, **89**, 5867–5872.

Aki, K. and Richards, P. G., 1980. *Quantitative Seismology, Theory and Methods*, W. H. Freeman, San Francisco .

Aki, K., Christofferson, A. and Husebye, E., 1977. Determination of three dimensional seismic structures of the lithosphere. *J. Geophys. Res.*, **82**, 277–296.

Al-Chalabi, M., 1979. Velocity determination from seismic reflection data. In Fitch, A. A. (ed.), *Developments in Geophysical Exploration Methods*, Vol. 1, Applied Science Publishers, London, pp. 1–68.

Ali, J. W. and Hill, I. A., 1991. Reflection seismics for shallow geological investigations: a case study from central England. *J. Geol. Soc. London*, **148**, 219–222.

Allegre, C. J., 1988. *The Behaviour of the Earth*, Harvard University Press.

Allen, C. R., 1965. *Phil. Trans. Roy. Soc.*, A, **258**, 82.

Allen, C. R., 1969. Active faulting in northern Turkey. *Div. Geol. Sciences*, California Inst. Technology, Pasadena, p. 32.

Allen, C. R., 1986. Seismological and paleoseismological techniques of research in active tectonics. In *Active Tectonics, Studies in Geophysics*, Nat. Academy Press, Washington, pp. 148–154.

Allen, R., 1982. Automatic phase pickers: their present and future prospects. *Bull. Seism. Soc. Am.*, **72**, 225–242.

Allen, S. J., 1980. Seismic method. *Geophysics*, **45**, 1619–1633.

Al-Sadi, H. N., 1980. *Seismic Exploration – Technique and Processing*, Birkhauser Verlag, Basle, 215 pp.

Alterman, Z., Zarosch, H. and Pekeris, C. L., 1959. Oscillations of the Earth. *Proc. Roy. Soc.*, A, **252**, 80.

Ambraseys, N. N., 1992. Long-term seismic hazard in the eastern Mediterranean region. In McCall, G. J. N., Laming, D. J. C. and Scott, S. C. (eds), *Geohazards: Natural and Man-Made*, Chapman and Hall, pp. 83–92.

Anderson, D. L., 1974. The interior of the Moon. *Physics Today*, **27**, 44–49.

Anderson, D. L., 1989. *Theory of the Earth*, Blackwell, London.

Anderson, D. L., 1995. Lithosphere, asthenosphere and perisphere. *Rev. Geophys.*, **33**, 125–149.

Anderson, D. L. and Dziewonski, A. M., 1984. Seismic tomography. *Sci. Am.*, **251**, 60–68.

Anderson, D. L. and Hart, R. S., 1976. An earth model based on free oscillations and body waves. *J. Geophys. Res.*, **81**, 1461–1475.

Anderson, H. and Webb, T., 1994. New Zealand seismicity: patterns revealed by the upgraded National Seismograph Network. *N.Z. J. Geol. Geophys.*, **37**, 477–493.

Anderson, J. G., 1991. Strong motion seismology. *Rev. Geophys.*, Suppl., US Nat. Rep. to IUGG, 1987–1990, 700–720.

Anderson, J. G., Bodin, P., Brune, J. N., Prince, J., Singh, S. K., Quass, R. and Onate, M., 1986. Strong ground motion from the Michoacan, Mexico, earthquake. *Science*, **233**, 1043–1049.

Anstey, N. A., 1977. *Seismic Interpretation*, IHRDC, Boston.

Arieh, E., 1994. Seismicity of Israel and adjacent areas. In Rutenberg, A. (ed.), *Earthquake Engineering*, Balkema, Rotterdam.

Argus, D. F. and Gordon, R. G., 1990. Pacific–North American plate motion from very long baseline interferometry compared with motion inferred from magnetic anomalies, transform faults and earthquake slip vectors. *J. Geophys. Res.*, **95** (B11), 17 315–17 324.

Arya, V. K. and Aggarwal, J. K. (eds), 1962. *Deconvolution of Seismic Data*, Hutchinson Ross Publishing.

Asquith, G. B. and Gibson, C. R., 1982. *Basic Well Logging for Geologists*, Am. Assoc. Petr. Geol., Tulsa, Oklahoma.

Atwater, B. F., Stuiver, M. and Yamaguchi, D. K., 1991. Radiocarbon test of earthquake magnitude at the Cascadia subduction zone. *Nature*, **353**, 156–158.

Backus, M. M., 1959. Water reverberations – their nature and elimination. *Geophysics*, **24**, 233–261.

Backus, M. M. and Chen, R. L., 1975. Flat spot exploration. *Geophys. Prospect.*, **23**, 533–577.

Badley, M. E., 1985. *Practical Seismic Interpretation*, Inter. Human Res. Dev. Corp., Boston.

Baeten, G. and Ziolkowski, A., 1990. *The Vibroseis Source*, Elsevier, Amsterdam.

Bak, P. and Tang, C., 1989. Earthquakes as a self-organised critical phenomena. *J. Geophys. Res.*, **94**, 15 635–15 637.

Bakun, W. H. and McEvilly, T. V., 1984. Recurrence models and Parkfield, California, earthquakes. *J. Geophys. Res.*, **89**, 3051–3058.

Bakun, W. H. and Lindh, A. G., 1985. The Parkfield, California, prediction experiment. *Science*, **229**, 619–624.

Balch, A. H. and Lee, M. W. (eds), 1984. *Vertical Seismic Profiling*, Inter. Human Res. Dev. Corp., Boston.

Barazangi, M. and Brown, L. (eds), 1986a. *Reflection Seismology: A Global Perspective*, Am. Geophys. Union, Geodynamic Series, 13.

Barazangi, M. and Brown, L. (eds), 1986b. *Reflection Seismology: The Continental Crust*, Am. Geophys. Union, Geodynamic Series, 14.

Barbier, M. G., 1983. *The Mini-Sosie Method*, Inter. Human Res. Dev. Corp., Boston.

Barka, A. A., 1992. The North Anatolian fault zone. *Annales Tectonicae*, **VI** (Suppl.), 164–195.

Basham, P. W. and Gregerson, S., 1989. Earthquakes at and near North Atlantic margins: neotectonics and post-glacial rebound – a workshop report. *Terra Nova*, **1**, 18–122.

Bates, C. C., Gaskell, T. F. and Rice, R. B., 1982. *Geophysics in the Affairs of Man*, Pergamon, Elmsford, NY.

Bath, M., 1967, Earthqaukes, large, destructive. In Runcorn, S. K. (ed.), *International Dictionary of Geophysis*, Pergamon Press, New York, 2 vols, 1728 pp.

Bayerly, M. and Brooks, M., 1980. A seismic study of deep structure in South Wales using quarry blasts. *Geophys. J. Roy. Astron. Soc.*, **60**, 1–19.

Becker, K., Sakai, H., Adamson, C. *et al.* (25 authors), 1989. Drilling deep into the young oceanic crust, Hole 504B, Costa Rica rift. *Rev. Geophys.*, **27**, 79–102.

Benioff, H., 1959. Fused-quartz extensometer for secular, tidal and seismic strains. *Bull. Geol. Soc. Am.*, **70**, 1019–1032.

Benioff, H., 1964. Earthquake source mechanisms *Science*, **143**, 1400.

Ben-Menahem, A. and Singh, S. J., 1981. *Seismic Waves and Sources*, Springer-Verlag, New York.

Berke, P. R. and Beatley, T., 1992. *Planning for Earthquakes*, Johns Hopkins University Press.

Berkhout, A. J., 1987. *Applied Seismic Wave Theory*, Elsevier, Amsterdam.

Beroza, G. C. and Jordan, T. H., 1990. Searching for slow and silent earthquakes using free oscillations. *J. Geophys. Res.*, **95**, 2485–2510.

Boatwright, J. and Choy, G. L., 1992, Acceleration source spectra anticipated for large earthquakes in N. E. North America. *Bull. Seism. Soc. Am.*, **82**, 660–682.

Bolt, B. A., 1976. *Nuclear Explosions and Earthquakes: The Parted Veil*, W. H. Freeman, New York.

Bolt, B. A., 1987. 50 years of studies on the inner core. *EOS, Trans. Am. Geophys. Un.*, **68**, 73.

Bolt, B. A., 1991. Balance of risks and benefits in preparation for earthquakes. *Science*, **251**, 169–174.

Bolt, B. A., 1993. *Earthquakes*, W. H. Freeman, San Francisco.

Boore, D. M., 1977. The motion of the ground in earthquakes. *Scientific American*, **237**, 68–78.

Boore, D. M., 1989. The Richter scale: its development and use for determining earthquake source parameters. *Tectonophysics*, **66**, 1–14.

Bott, M. H. P., 1982. *The Interior of the Earth*, 2nd edn, Edward Arnold, London.

Brady, B. T. and Rowell, B. T., 1986. Laboratory investigation of the electrodynamics of rock fracture. *Nature*, **321**, 488–492.

Braile, L. W., 1991. Seismic studies of the Earth's crust. *Rev. Geophys.*, (Suppl.), US Nat. Report to IUGG 1987–1990, 680–687.

Brocher, T. M. and Hart, P. E., 1991. Comparison of vibroseis and explosive source methods for deep crustal seismic reflection profiling in the Basin and Range Province. *J. Geophys. Res.*, **96** (B11), 18 197–18 213.

Brown, J. M. and McQueen, R. G., 1983. The equation of state for iron and the Earth's core. In Akimoto, S. and Manghnani, M. (eds), *High Pressure Research in Geophysics, Adv. Earth and Planet. Sci.*, **12**, 611–623. Centre for Academic Publishing, Tokyo.

Brown, L. F. and Fisher, W. L., 1980. *Seismic Stratigraphic Interpretation and Petroleum Exploration*, Am. Assoc. Petrol. Geol., Continuing Education Course Note Series No. 16.

Brown, L., Wille, D., Zheng, L. *et al.* (14 others), 1987. COCORP: New perspectives on the deep crust. *Geophys. J. Roy. Astron. Soc.*, **89**, 47.

Brown, R. D., Wallace, R. E. and Hill, D. P., 1992. The San Andreas fault system, California, USA. *Annales Tectonicae*, **VI** (Suppl.), 261–284.

Brune, J. N., 1970. Tectonic stress and the spectra of seismic shear waves from earthquakes. *J. Geophys. Res.*, **75**, 4997–5009.

Brune, J. N., 1991. Seismic source dynamics, radiation and stress. *Rev. Geophys.*, (Suppl.), US Nat. Report to IUGG, pp. 688–699.

Brune, J. N. and Oliver, J., 1959. The seismic noise of the Earth's surface. *Bull. Seism. Soc. Am.*, **49**, 349–353.

Brush, S. G., 1980. Discovery of the Earth's core. *Am. J. Phys.*, **48**, 705–720.

Buland, R., 1976. The mechanics of locating earthquakes. *Bull. Seism. Soc. Am.*, **66**, 173–187.

Bullen, K. E., 1975. *The Earth's Density*, Chapman and Hall, London.

Bullen, K. E. and Bolt, B. A., 1985. *An Introduction to the Theory of Seismology*, 4th edn, Cambridge University Press.

Burger, H. R., 1992, *Exploration Geophysics of the Shallow Subsurface*, Prentice-Hall, Englewood Cliffs, NJ.

Cagniard, L., 1939. *Reflexion et Refraction des Ondes Seismiques Progressives*, Gauthier-Villars and Cie, Paris.

Calvert, A. J., 1990. Ray-tracing based prediction and subtraction of water-layer multiples. *Geophysics*, **55**, 443–451.

Cao, T. and Aki, K., 1983. Assigning probability gain for precursors of four large Chinese earthquakes. *J. Geophys. Res.*, **88** (B3), 2185–2190.

Carder, D. S., Gordon, D. W. and Jordan, J. N., 1966. Analysis of surface foci travel-times. *Bull. Seism. Soc. Am.*, **56**, 815–840.

Carter, J. A., Barstow, N., Pomeroy, P. W., Chael, E. P. and Leahy, P. J., 1991. High frequency seismic noise as a function of depth. *Bull. Seism. Soc. Am.*, **81**(4), 1101–1114.

Cassell, B., 1984. Vertical seismic profiles — an introduction. *First Break*, **12**, 9–18.

Cerveny, V. and Ravindra, R., 1971. *Theory of Seismic Head Waves*, University of Toronto Press, Toronto.

Cerveny, V., Molotkov, I. A. and Psencik, I., 1977. *Ray Method in Seismology*, University of Karlova, Prague.

Chapman, C. H., 1978. A new method for computing synthetic seismograms. *Geophys. J. Roy. Astron. Soc.*, **54**, 431–518.

Chen, W. P. and Molnar, P., 1983. Focal depths of intracontinental and intraplate earthquakes and their implications for the thermal and mechanical properties of the lithosphere. *J. Geophys. Res.*, **88**, 4183–4214.

Chen Yong, Kam-Ling Tsoi, Chen Feibi, Gao Zhenhuan, Zou Qijia and Chen Zhangli, 1988. *The Great Tangshan Earthquake of 1976: An Anatomy of Disaster*, State Seismological Bureau, China, Pergamon Press.

Chester, D., 1993. *Volcanoes and Society*, Edward Arnold.

Chester, F. M., Evans, J. P. and Biegel, R. L., 1993. Internal structure and weakening mechanisms of the San Andreas fault. *J. Geophys. Res.*, **98** (B1), 771–786.

Choy, G. L., 1990. Global digital seismic data: interpretation of the earthquake mechanism from broadband data. Pacific Rim 90 Congress, *Australasian Inst. Min. and Metall.*, **295**, 11–18.

Christensen, N. I., 1989. Seismic velocities. In Carmichael, R. S. (ed.), *Practical Handbook of Physical Properties of Rocks and Minerals*, CRC Press, Boca Raton, Florida, pp. 429–546.

Christensen, N. I. and Salisbury, M. H., 1975. Structure and constitution of the lower oceanic crust. *Rev. Geophys.*, **13**, 57–86.

Christensen, N. I. and Szymanski, D. L., 1991. Seismic properties and the origin of reflectivity from a classic Paleozoic sedimentary sequence, Valley and Ridge province, southern Appalachians. *Geol. Soc. Am. Bull.*, **103**, 277–289.

Christensen, N. I. and Wefner, W. W., 1989. Laboratory techniques for determining seismic velocities and attenuations, with applications to continental lithosphere. In Pakiser, L. C. and Mooney, W. D. (eds), *Geophysical Framework of the United States*, *Geol. Soc. USA Mem.*, **172**, 91–102.

Christensen-Dalsgaard, J., Gough, D. and Toomre, J., 1985. Seismology of the Sun. *Science*, **229**, 923–931.

Chun, J. H. and Jacewitz, C. A., 1981. Fundamentals of frequency domain migration. *Geophys*, **46**, 717–733.

Claerbout, J. F., 1985. *Imaging the Earth's Interior*, Blackwell, Palo Alto.

Clark, R. B., 1985. Enders Robinson. *Geophysics: The Leading Edge*, **4**, 16–20.

Clark, S. P. (ed.), 1966. *Handbook of Geophysical Constants*, Geol. Soc. Am. Memoir, 97.

Clark, I. F. and Cook, B. J. (eds), 1983. *Perspectives of the Earth*, Aust. Acad. Science, Canberra, 651 pp.

Clay, C. S., 1990. *Elementary Exploration Seismology*, Prentice-Hall, Englewood Cliffs, NJ.

Cleary, J. and Hales, A. L., 1966. An analysis of the travel times of P waves to North American stations, in the distance range 32° to 100°. *Bull. Seism. Soc. Am.*, **56**, 467–489.

Coburn, A. W. and Spence, R. J. S., 1992. *Earthquake Protection*, John Wiley, Chichester.

Coffeen, J. A., 1984. *Interpreting Seismic Data*, PennWell Books, Tulsa, OK.

Coffeen, J. A., 1986. *Seismic Exploration Fundamentals*, 2nd edn, PennWell Books, Tulsa, OK.

Coffeen, J. A., 1990. *Seismic on Screen*, PennWell Books, Tulsa, OK.

Connelly, D. L., Ferris, B. J. and Trembly, L. D., 1991. Northwestern Williston Basin case histories with 3D seismic data. *Geophysics*, **56**, 1849–1874.

Cordier, J. P., 1985. *Velocities in Reflection Seismology*, Reidel Publishing Co., Boston, MA.

Cormier, V. F., 1982. The effect of attenuation on seismic body waves. *Bull. Seism. Soc. Am.*, **72**, S169–200.

Cox, A. and Hart, R. B., 1986. *Plate Tectonics*, Blackwell Scientific, Palo Alto, CA.

Crampin, S., 1984. Effective anisotropic elastic constants for wave propagation through cracked solids. *Geophys. J. Roy. Astron. Soc.*, **76**, 135–145.

Crampin, S. and Lovell, J. H., 1991. A decade of shear-wave splitting in the Earth's crust: what does it mean? what use can we make of it? and what should we do next? *Geophys. J. Int.*, **107**, 387–407.

Cranswick, E., Gardner, B., Hammond, S. and Banfill, R., 1993. Recording ground motions where people live. *EOS, Trans. Am. Geophys. Un.*, **74**, 243–244.

Creager, K. C., 1992. Anisotropy of the inner core from differential travel times of the phases PKP and PKIKP. *Nature*, **356**, 309–314.

Creager, K. C. and Jordan, T. H., 1986. Slab penetration into the lower mantle. *J. Geophys. Res.*, **89**, 3031–3049.

Cross, T. A. and Lessenger, M. A., 1988. Seismic stratigraphy. *Ann. Rev. Earth and Planet. Sci.*, **16**, 319–354.

Dahle, A., Gjoystdal, H., Grammeltvedt, G. and Hansen, T. S., 1985. Application of seismic reflection methods for ore prospecting in crystalline rock. *First Break*, **3**, 9–16.

Dahlman, O. and Israelson, H., 1987. *Monitoring Underground Nuclear Explosions*, Elsevier, New York.

Dahm, C. G. and Graebner, R. J., 1982. Field development with three-dimensional seismic methods in the Gulf of Thailand — a case history. *Geophysics*, **47**, 149–176.

Danbom, S. H. and Domenico, S. N. (eds), 1986. *Shear Wave Exploration, Geophysical Developments No. 1*. Soc. Explor. Geophysicists, Tulsa, OK.

Das, S., Boatwright, J. and Scholz, C. H. (eds), 1986. *Earthquake Source Mechanisms*, Am. Geophys. Union Mon. 37, Maurice Ewing Series 6.

Davis, T. L. and Namson, J. S., 1994. A balanced cross-section of the 1994 Northridge earthquake, Southern California. *Nature*, **372**, 167–169.

Davison, C., 1927. *The Founders of Seismology*, Cambridge University Press, Cambridge.

Dawson, J. B., Carswell, D. A., Hall, J. and Wedepohl, J. H. (eds), 1986. *The Nature of the Lower Continental Crust*, Geological Society (UK), Publication No. 24.

Decker, R. W., 1986. Forecasting volcanic eruptions. *Ann. Rev. Earth and Planet. Sci.*, **14**, 267–291.

DeMets, C., Gordan, R. G., Argus, D. F. and Stein, S., 1990. Current plate motions. *Geophys. J.*, **101**, 425–478.

Denham, D., 1988. Australian seismicity — the puzzle of the not-so-stable continent. *Seism. Res. Lett.*, **59**, 235–240.

Der, Z. A., Rivers, W. D., McElfresh, T. W., O'Donnell, A., Klouda, P. J. and Marshall, M. E., 1982. Worldwide variations in the attenuative properties of the upper mantle as determined by spectral studies of short period body waves. *Phys. Earth and Planet. Interiors.*, **30**, 12–25.

Dewey, J. and Byerly, P., 1969. The early history of seismometry (to 1900). *Bull. Seism. Soc. Am.*, **59**, 183–227.

Dikkers, A. J., 1985. *Geology in Petroleum Production*, Elsevier, Amsterdam. Ch. F. 239 pp.

Dix, C. H., 1955. Seismic velocities from surface measurements. *Geophysics*, **20**, 68–86.

Dix, C. H., 1981. *Seismic Prospecting for Oil*, revised edn, MA. Int. Human Res. Dev. Corp, Boston.

Dobrin, M. B., 1976. *Itroduction to Geophysical Prospecting*, McGraw-Hill, New York.

Dobrin, M. B. and Savit, C. H., 1988. *Introduction to Geophysical Prospecting*, 4th edn, McGraw-Hill, London.

Dohr, G. P. (ed.), 1985. *Seismic Shear Waves*, Geophysical Press, London.

Domenico, S. N., 1984. Rock lithology and porosity determination from shear and compressional wave velocity. *Geophysics*, **49**, 1188–1195.

Dooley, J. C. (ed.), 1990. Proceedings of engineering workshop. *Explor. Geophys.*, **21**, 1–143.

Dorman, J., Ewing, M. and Oliver, J., 1960. Study of shear-wave velocity distribution in the upper mantle by mantle Rayleigh waves. *Bull. Seism. Soc. Am.*, **50**, 87–115.

Doyle, H. A., 1957. Seismic recordings of atomic explosions in Australia. *Nature*, **180**, 132–134.

Doyle, H. A., 1984. Earthquake prediction: present trends and the Australian situation. *Search*, **15**, 263–270.

Doyle, H. A. and Hales, A. L., 1967. An analysis of the travel times of S waves to North American stations, in the distance range 28° to 82°. *Bull. Seism. Soc. Am.*, **57**, 761–771.

Drummond, B. J. (ed.), 1991. *The Australian Lithosphere*, Geol. Soc. Austral., Spec. Pub., 17.

Dziewonski, A. M., 1989. Earth structure, global. In James, D. E. (ed.), *The Encyclopedia of Solid Earth Geophysics*, Van Nostrand Reinhold, New York, pp. 331–359.

Dziewonski, A. M. and Anderson, D. L., 1981. Preliminary reference Earth model. *Phys. Earth and Planet. Int.*, **25**, 297–356.

Dziewonski, A. M. and Anderson, D. L., 1983. Travel times and station corrections for P waves at teleseismic distances. *J. Geophys. Res.*, **88**, 3295–3314.

Dziewonski, A. M. and Anderson, D. L., 1984. Seismic tomography of the Earth's interior. *American Scientist*, **72**, 483–494.

Dziewonski, A. M. and Woodhouse, J. H., 1987. Global images of the Earth's interior, *Science*, **236**, 38–47.

Dziewonski, A. M., Ekstrom, G., Franzen, J. E. and Woodhouse, J. H., 1987. Global seismicity of 1977: centroid-moment tensor solutions for 524 earthquakes. *Phys. Earth and Planet. Int.*, **45**, 11–36.

Eiby, G. A., 1980. *Earthquakes*, Heinemann, London.

Einarsson, P., 1991. Earthquakes and present-day tectonism in Iceland. *Tectonophysics*, **189**, 261–279.

Evenden, B. S. and Stone, D. R., 1971. *Seismic Prospecting Instruments, Vol. 2, Instrument Performance and Testing*, Borntraeger, Berlin.

Ewing, M., 1965, The sediments of the Argentine basin. *Q. J. Roy. Astron. Soc.*, **6**, 10–27.

Ewing, M. and Press, F., 1950. Crustal structure and surface wave dispersion. *Bull. Seism. Soc. Am.*, **40**, 271–280.

Ewing, M., Jardetsky, W. S. and Press, F., 1957. *Elastic Waves in Layered Media*, McGraw-Hill, London.

Fairhurst, C. (ed.), 1990. *Rockbursts and Seismicity in Mines*, Balkema, Rotterdam.

Fehler, M., 1985. Locations and spectral properties of earthquakes accompanying an eruption of Mt St Helens. *J. Geophys. Res.*, **90**, 12 729–12 740.

Field, E. H., Hough, S. H. and Jacob, A. H., 1990. Using microtremors to assess potential earthquake site response: a case study in Flushing Meadows, New York. *Bull. Seism. Soc. Am.*, **80**, 1456–1480.

Finlayson, D. M., 1968. First arrival data from the Carpentaria region upper mantle project (CRUMP). *J. Geol. Soc. Aust.*, **15**(1), 35–50.

Finlayson, D. M., Wake-Dyster, K. D., Leven, J. H., Johnstone, D. W., Murray, C. G., Harrington, H. J., Korsch, R. J. and Wellman, P., 1990. Seismic imaging of major tectonic features in the crust of Phanerozoic eastern Australia. *Tectonophysics*, **173**, 211–230.

Fitch, A., 1979. Earthquakes and plate tectonic s Ch. 15 in McIlkinny, M. W. (ed.), *The Earth, its Origin, Structure and Evolution*, Academic Press, London.

Fix, J. E., 1972. Ambient earth motion in the period range from 0.1 to 2560 seconds. *Bull. Seism. Soc. Am.*, **62**, 1753–1760.

Fountain, D. M. and Christensen, N. I., 1989. Constitution of the continental crust and upper mantle: a review. In Pakiser, L. C. and Mooney, W. D. (eds), *Geophysical Framework of the Continental United States*, Geol. Soc. Amer., Mem. 172, pp. 711–742.

Franchetau, J., 1983. The oceanic crust. *Scientific American*, **249** (3), 68–84.

Frankel, A., 1991. Mechanisms of seismic attenuation in the crust: scattering and anelasticity in New York State, South Africa and Southern California. *J. Geophys. Res.*, **96** (B4), 6269–6289.

Frohlich, C., 1987. Kiyoo Wadati and early research on deep focus earthquakes: introduction to special section on deep and intermediate focus earthquakes. *J. Geophys. Res.*, **92**, 13 777–13 788.

Frohlich, C., 1989. The nature of deep-focus earthquakes. *Ann. Rev. Earth and Planet. Sci.*, **17**, 227–254.

Frohlich, C., 1994a. A break in the deep. *Nature*, **368**, 100–101.

Frohlich, C., 1994b. Earthquakes with non-double-couple mechanisms. *Science*, **264**, 804–809.

Frohlich, C. and Davis, S. C., 1993. Teleseismic b values; or, much ado about 1.0. *J. Geophys. Res.*, **98** (B1), 631–644.

Fuchs, K. and Froidevaux, C., 1987. *Composition, Structure and Dynamics of the Lithosphere–Asthenosphere System*, Am. Geophys. Un., Geodynamics Series, 16.

Fuchs, K., Kovlovsky, Y. A., Krivtsov, A. I. and Zoback, M. D., 1990. *Super-Deep Continental Drilling and Deep Geophysical Sounding*, Springer-Verlag, Berlin.

Gadallah, M. R., 1994. *Reservoir Seismology*, PennWell Books, Tusla, OK.

Gardner, G. H. F., 1985. *Migration of Seismic Data*, Soc. Explor. Geophys., Geophysics Reprint Series, 4. Soc. Explor. Geophys., Tulsa, OK.

Gardner, G. H. F., Gardner, L. W. and Gregory, A. R., 1974. Formation velocity and density — the diagnostic basics for stratigraphic traps. *Geophysics*, **39**, 770–780.

Garland, G. D., 1979. *Introduction to Geophysics*, ch. 7, Saunders, Philadelphia, PA.

Giardini, D., 1988. Frequency distribution and quantification of deep earthquakes, *J. Geophys. Res.*, **93**, 2095–2105.

Gibowicz, S. J., 1990. Seismicity induced by mining. *Advances in Geophysics*, **32**, 1–74. Academic Press, San Diego, CA.

Gladwin, M. T., Gwyther, R. L., Higbie, J. W. and Hart, R. G., 1991. A medium term precursor to the Loma Prieta earthquake? *Geophys. Res. Lett.*, **18**, 1377–1380.

Gibson, B., Larner, K. and Levin, S., 1983. Efficient 3-D migration in two steps. *Geophys. Prospect.*, **31**, 1–33.

Goleby, B. R., Shaw, R. D., Wright, C., Kennett, B. L. N. and Lambeck, K., 1989. Geophysical evidence for 'thick-skinned' crustal deformation in central Australia. *Nature*, **337**, 325.

Gordon, F. R. and Lewis, J. D., 1980. *The Meckering and Calingiri Earthquakes, October 1968 and March 1970*, Bull. Geol. Surv. Western Australia, 126.

Gordon, R. G. and Stein, S., 1992. Global tectonics and space geodesy. *Science*, **256**, 333–342.

Goupillaud, P. L., 1976. Signal design in the Vibroseis technique. *Geophysics*, **41**, 1291–1304.

Grand, S. P., 1994. Mantle shear structure beneath the Americas and surrounding oceans. *J. Geophys. Res.*, **99** (B6), 11 591–11 621.

Grant F. S. and West, G. F., 1965. *Interpretation Theory in Applied Geophysics*, McGraw-Hill, New York.

Grantz, A. (ed.), 1971. *The San Fernando, California, Earthquake of February 9, 1971*, US Geol. Surv. Professional Paper 733.

Grau, G. and Lailly, P., 1989. Seismic wavefield migration. In James, D. E. (ed.), *The Encyclopedia of Solid Earth Geophysics*, Van Nostrand Reinhold, New York, pp. 1151–1166.

Green, H. W. II, 1994. Solving the paradox of deep earthquakes. *Scientific American*, Sept., 50–57.

Green, R. W. E. and Hales, A. L., 1968. The travel times of P waves to 30° in the central United States and upper mantle structure. *Bull. Seism. Soc. Am.*, **58**, 267–289.

Gregersen, S., Korhonen, H. and Husebye, E. S., 1991. Fennoscandian dynamics: present-day earthquake activity. *Tectonophysics*, **189**, 333–344.

Griffin, W. L. and O'Reilly, S. Y., 1987. Is the continental Moho the crust–mantle boundary? *Geology*, **15**, 241–244.

Griffiths, D. H. and King, R. F., 1965. *Applied Geophysics for Engineers and Geologists*, ch. 4, Pergamon Press, Oxford.

Gubbins, D., 1990. *Seismology and Plate Tectonics*, Cambridge University Press, Cambridge.

Gutenberg, B., 1924. Der Aufbau der Erdkruste auf Grund geophysikalischer Bertrachtungen, *Z. Geophys.*, **1**, 94–108.

Gutenberg, B., 1956. The energy of earthquakes. *Quart. J. Geol. Soc. London*, **112**, 1–14.

Gutenberg, B., 1959. *Physics of the Earth's Interior*, Academic Press, New York.

Gutenberg, B. and Richter, C. F., 1954. *Seismicity of the Earth*, 2nd edn, Princeton University Press.

Gutenberg, B. and Richter, C. F., 1956. Earthquake magnitude, intensity, energy and acceleration. *Bull. Seism. Soc. Am.*, **46**, 105–143.

Hagedoorn, J. G., 1954. A process of seismic reflection interpretation. *Geophys. Prospect.*, **21**, 85–127.

Hager, B. H. and O'Connell, R. J., 1979. Kinematic models of large-scale flow in the Earth's mantle. *J. Geophys. Res.*, **84**, B3, 1031–1048.

Hager, B. H., King, R. W. and Murray, M. H., 1991. Measurement of crustal deformation using the global positioning system. *Ann. Rev. Earth Planet. Sci.*, **19**, 351–382.

Hajnal, Z., Stauffer, M. R., King, M. S., Wallis, P. F., Wang, H. F. and Jones, L. E. A., 1983. Seismic characteristics of a Precambrian pluton and its adjacent rocks. *Geophysics*, **48**, 569–581.

Hales, A. L., 1991. Upper mantle models and the thickness of the continental lithosphere. *Geophys. J. Int.*, **105**, 355–363.

Hales, A. L. and Doyle, H. A., 1967. P and S travel time anomalies and their interpretation. *Geophys. J. Roy. Astron. Soc.*, **13**, 403–415.

Hanks, T. C. and Kanamori, H., 1979. A moment magnitude scale. *J. Geophys. Res.*, **84**, 2348–2350.

Hanks, T. C. and Krawinkler, H., 1991. The 1989 Loma Prieta earthquake and its effects: introduction to the special issue. *Bull. Seism. Soc. Am.*, **81**, 1415–1423.

Hanks, T. and Wyss, M., 1972. The use of body-wave spectra in the determination of seismic-source parameters. *Bul. Seism. Soc. Am.*, **62**, 561–589.

Hardage, B. A., 1985. Vertical seismic profiling: Part A – principles. In Helbig, K. and Treitel, S. (eds), *Handbook of Geophysical Exploration*, Vol. 9, Geophysical Press, Netherlands.

Hardage, B. A. and Toksoz, M. N., 1983. *Vertical Seismic Profiling, Parts A and B*, Geophysical Press, The Netherlands.

Hardy, R. J. J. and Hobbs, R. W., 1991. A strategy for multiple suppression. *First Break*, **9**, 139–144.

Hasegawa, A., Horiuchi, S. and Umino, N., 1994, Seismic structure of the northeastern Japan convergent margin: a synthesis. *J. Geophys. Res.*, **99**, B11, 22295–22311.

Hatherly, P. J. and Neville, M. J., 1986. Experience with the generalized reciprocal method of seismic refraction interpretation for shallow engineering site evaluation. *Geophysics*, **51**, 255–265.

Hatton, L., Worthington, M. H. and Makin, J., 1986. *Seismic Data Processing*, Blackwell Scientific Pubs. Oxford.

Hawkins, L. V. and Maggs, D., 1961. Nomograms for determining maximum errors and limiting conditions in seismic refraction survey with a blind zone problem. *Geophys. Prospect.*, **9**, 526–533.

Hawkins, L. V. and Whiteley, R. J., 1981. Shallow seismic refraction survey of the Woodlawn orebody. In Whiteley, R. J. (ed.). *Geophysical Case Study of the Woodlawn Orebody, New South Wales, Australia*, Pergamon, Oxford, pp. 497–506.

Heaton, T. H. and Hartzell, S. H., 1988. Earthquake ground motions. *Ann. Rev. Earth and Planet. Sci*, **16**, 121–145.

Helbig, K. and Treitel, S. (eds), 1987. Seismic stratigraphy. In Hardage, B. A. (ed.), *Handbook of Geophysical Exploration*, Vol. 9.

Herrin, E. and Taggart, J., 1962. Regional variations in Pn velocity and their effect on the location of epicentres. *Bull. Seism. Soc. Am.*, **52**. 1037–1046.

Herrmann, R. B., 1975. A student's guide to the use of P and S wave data for focal mechanism determination. *Earthquake Notes*, **46**, 21–39.

Hileman, J. A., Embree, P. and Pflueger, J. C., 1968. Automated static corrections. *Geophys. Prospect.*, **16**, 326–358.

Hill, D. R., Wallace, R. E. and Cockerham, R. S., 1985. Review of evidence on the potential for major earthquakes and volcanism in the Long Valley–Mono craters–White Mt. region of eastern California. *Earthq. Prediction Res.*, **3**, 571–594.

Hisamoto, S. and Murayama, C., 1961. Drawing the wave-fronts of the Chilean tsunami of May 23, 1960. *Quart. J. Seismol.*, **26**, 1 (in Japanese).

Hokkaido Tsunami Survey Group, 1993. Tsunami devastates Japanese coastal region. *EOS, Trans. Am. Geophys. Un.*, **74**, 417–432.

Holbrook, W. S., Mooney, W. D. and Christensen, N. I., 1992. The seismic velocity structure of the deep continental crust. In Fountain, D. M., Arculus, R. and Kay, R. W. (eds), *Continental Lower Crust*, Elsevier, Amsterdam, pp. 1–43.

Hood, P., 1979. Migration. In Fitch, A. A. (ed.), *Developments in Geophysical Exploration Methods — 2*, Applied Science Publishers, London, pp. 151–230.

Hopkins, W., 1839. Researches in physical geology. *Phil. Trans. Roy. Soc. London*, A, **129**, 381.

Howell, B. F. Jr., 1986. History of ideas on the cause of earthquakes. *EOS, Trans. Am. Geophys. Un.*, **67**, 1323–1326.

Howell, B. F. Jr., 1989. Seismic instrumentation: history. In James, D. E. (ed.), *The Encyclopedia of Solid Earth Geophysics*, Van Nostrand Reinhold, New York, pp. 1037–1044.

Howell, B. F. Jr., 1990. *An Introduction to Seismological Research: History and Development*, Cambridge University Press, Cambridge.

Huang, H., Spencer, C. and Green, A. 1986. A method for inversion of refraction and reflection travel times for laterally varying velocity structures. *Bull. Seism. Soc. Am.*, **76**, 837–846.

Hunter, J. A., Pullan, S. E., Burns, R. A., Gagne, R. M. and Good, R. L., 1984. Shallow seismic reflection mapping of the overburden–bedrock interface with the engineering seismograph. Some simple techniques. *Geophysics*, **49**, 1381–1385.

Hyndman, R. D., 1988. Dipping seismic reflectors, electrically conducting zones, and trapped water in the crust over a subducting plate. *J. Geophys. Res.*, **93**, 13 391–13 405.

Ihmle, P. F., Harabaglia, P. and Jordan, T. H., 1993.

Teleseismic detection of a slow precursor to the great 1989 Macquarie Ridge earthquake. *Science*, **261**, 177–183.

Iida, K. and Iwasaki, T. (eds), 1983. *Tsunamis: Their Science and Engineering*, D. Reidel, Boston.

Ikwuakor, K. C., 1988. V_p/V_s revisited: pitfalls and new interpretation techniques. *World Oil*, Sept., 41–46.

Isacks, B., Oliver, J. and Sykes, L. R., 1968. Seismology and the new global tectonics. *J. Geophys. Res.*, **73**, 5855–5899.

Iyer, H. M., 1970. Seismology. In Tucker, R. H., Cook, A. H., Iyer, H. M. and Stacey, F. D. (eds), *Global Geophysics*, English Universities Press.

Iyer, H. M., 1984. Geophysical evidence for the locations, shapes and sizes, and internal structures of magma chambers beneath regions of Quaternary volcanism. *Roy. Soc. London Phil. Trans.*, **310**, 473–510.

Iyer, H. M. and Hirahara, K. (eds), 1993. *Seismic Tomography*, Chapman and Hall, London.

Iyer, H. M. and Hitchcock, T., 1989. Upper-mantle structure in the continental US and Canada. In Pakiser, L. C. and Mooney, W. D. (eds), *Geophysical Framework of the Continental United States*, Geol. Soc. Memoir No. 172, pp. 681–710.

Jacob, J. H. and Turkstra, C. J. (eds), 1989. *Earthquake Hazards and the Design of Constructed Facilities in the Eastern USA*, Annals of the New York, Academy of Sci., 558. New York Academy of Science.

Jacobs, J., 1987. *The Earth's Core*, Academic Press, New York.

Jacobson, R. S., Dorman, L. M., Purdy, G. M., Schultz, A. and Solomon, S. C., 1991. Ocean bottom facilities available. *EOS, Trans. Am. Geophys. Un.*, **72**, 506–515.

Jain, K. C. and de Figueiredo, R. J. P. (eds), 1982. *Concepts and Techniques in Oil and Gas Exploration*, Soc. Explor. Geophys., Tulsa, OK.

Jarchow, C. M. and Thompson, G. A., 1989. The nature of the Mohorovicic Discontinuity. *Ann. Rev. Earth and Planet. Sci.*, **17**, 475–506.

Jeanloz, R., 1988. High-pressure experiments and the earth's deep interior. *Physics Today*, S44–45.

Jeanloz, R., 1990. The nature of the Earth's core. *Ann. Rev. Earth and Planet. Sci.*, **18**, 357–386.

Jeanloz, R. and Lay, T., 1993. The core–mantle boundary. *Scientific American*, **268** (5), 26–33.

Jeffreys, H., 1926a. The rigidity of the earth's central core. *Mon. Not. Roy. Astron. Soc. Geophys.*, Suppl., **1**, 371–383.

Jeffreys, H., 1926b. On compressional waves in two superposed layers. *Proc. Cambridge Phil. Soc.*, **22**, 472–481.

Jeffreys, H., 1970. *The Earth*, 5th edn. Cambridge University Press, London.

Jeffreys, H. and Bullen, K. E., 1940. *Seismological Tables*, British Assoc. Advancement of Sci., Gray-Milne Trust, London.

Johnson, J. D. and French, W. S., 1992. Migration — the inverse method. In Jain, K. C. and de Figueiredo, R. J. P. (eds), *Concepts and Techniques in Oil and Gas Exploration*, Soc. Exploratory Geophysics, Tulsa, OK, p.126.

Johnston, A. C., 1991. Light from seismic sources. *Nature*, **354**, 361.

Johnston, A. C. and Kanter, L. R., 1990. Earthquakes in stable continental crust. *Scientific American*, March, pp. 42–49.

Johnston, J. E. and Christensen, N. I., 1992. Shear wave reflectivity, anisotropies, Poisson's ratios, and densities of a southern Appalachian Paleozoic sedimentary sequence. *Tectonophysics*, **210**, 1–20.

Johnston, M. J. S., 1989. Review of magnetic and electric field effects near active faults and volcanoes in the USA. *Phys. Earth and Planet. Int.*, **57**, 47–63.

Jones, L. M., 1985. Foreshocks and time-dependent earthquake hazard assessment in southern California. *Bull. Seism. Soc. Am.*, **75**, 1669–1680.

Jordan, T. H., Lerner-Lam, A. L. and Creager, K. C., 1989. Seismic imaging of boundary layers and deep mantle convection. In Peltier, W. R. (ed.), *Mantle Convection: Plate Tectonics and Global Dynamics*, Gordon and Breach, New York, pp. 97–201.

Joyce, C., 1991. Standing up to earthquakes. *New Scientist*, **129**(1756), 22–27.

Joyner, W. B. and Boore, D. M., 1981. Peak horizontal acceleration and velocity from strong-motion records including records from the 1979 Imperial Valley, California, earthquake. *Bull. Seismol. Soc. Amer.*, **71**, 2011–2038.

Juhasova, E., 1991. *Seismic Effects on Structures*, Elsevier, Amsterdam.

Julian, B. R. and Anderson, D. L., 1968. Travel times, apparent velocities and amplitudes of body waves. *Bull. Seismol. Soc. Amer.*, **58**, 339–366.

Kagan, Y. Y. and Jackson, D. D., 1995. New seismic gap hypothesis: five years after. *J. Geophys. Res.*, **100**, B3, 3943–3959.

Kanamori, H., 1977. The energy release in great earthquakes. *J. Geophys. Res.*, **82**, 2981–2987.

Kanamori, H., 1983. Magnitude scale and the quantification of earthquakes. *Tectonophysics*, **93**, 185–199.

Kanamori, H., 1994. Mechanics of earthquakes. *Ann. Rev. Earth and Planet. Sci.*, **22**, 207–237.

Kanamori, H. and Kikuchi, M., 1993. The 1992 Nicaraguan earthquake: a slow tsunami earthquake associated with subducted sediments. *Nature*, **361**, 714–716.

Kanasewich, E., 1990. *Seismic Noise Attenuation*, Vol. 7, Seismic Exploration Series, Pergamon Press, Oxford.

Kappus, M. E., Harding, A. J. and Orcutt, J. A., 1990. A comparison of tau-p transform methods. *Geophysics*, **55** (9), 1202–1215.

Karato, S. and Spetzler, H. A., 1990. Defect microdynamics in minerals and solid-state mechanisms of seismic wave attenuation and velocity dispersion in the mantle. *Rev. Geophys.*, **28**, 399–421.

Kaufman, H., 1953. Velocity functions in seismic prospecting. *Geophysics*, **18**, 289–297.

Kearey, P. and Brooks, M., 1991. *An Introduction to Geophysical Exploration*, Oxford University Press.

Kearey, P. and Vine, F., 1990. *Global Tectonics*, Blackwell, Oxford.

Kennett, B. L. N., 1977. Towards a more detailed seismic picture of the oceanic crust and mantle. *Marine Geophysical Researches*, **3**, 7–42.

Kennett, B. L. N., 1983. *Seismic Wave Propagation in Stratified Media*, Cambridge University Press, London.

Kennett, B. L. N. and Engdahl, E. R., 1991. Traveltimes for global earthquake location and phase identification. *Geophys. J. Int.*, **105**, 429–465.

Kennett, B. L. N. and Harding, A. J., 1985. Is ray theory adequate for reflection seismic modelling (a survey of modelling methods). *First Break*, **3** (1), 9–14.

Kennett, P., Ireson, R. L. and Conn, P. J., 1980. Vertical seismic profiles: their application in exploration geophysics. *Geophys. Prospect.*, **28**, 676–699.

Kent, G. M., Harding, A. J. and Orcutt, A. J., 1990. Evidence for a smaller magma chamber beneath East Pacific Rise at 9° 30′ N, *Nature*, **344**, 650–653.

Key, D., 1988. *Earthquake Design Practice for Buildings*, Thomas Telford, London.

King, C.-Y., 1986. Gas geochemistry applied to earthquake prediction: an overview. *J. Geophys. Res.*, **91**, 12 269–12 281.

Kirby, S. H., Durham, W. B. and Stern, L. A., 1991. Mantle phase changes and deep-earthquake faulting in subducting lithosphere. *Science*, **252**, 216–225.

Klemperer, S. L. and Hobbs, R. (eds), 1992. *The BIRPS Atlas*, Cambridge University Press, Cambridge.

Kleyn, A. H., 1983. *Seismic Reflection Interpretation*, Applied Science Publishers, London.

Knittle, E. and Jeanloz, R., 1991. Earth's core–mantle boundary: results of experiments at high pressures and temperatures. *Science*, **251**, 1438–1443.

Knott, C. G., 1899. Reflexion and refraction of elastic waves, with seismological applications. *Phil. Mag.*, **48**, (5), 64–97, 567–569.

Kovach, R. L., 1978. Seismic surface waves and crustal and upper mantle structure. *Rev. Geophys. and Space Phys.*, **16**, 1–13.

Kozlovsky, Y. A. (ed.), 1987. *The Super-Deep Well of the Kola Peninsula*, Springer-Verlag, Berlin.

Krapp, R. W., 1990. Vertical resolution of thick beds, thin beds, and thin-bed cyclothems. *Geophysics*, **55**, 1183–1190.

Krinitzsky, E. L., Gould, J. P. and Edinger, P. H., 1993. *Fundamentals of Earthquake Resistant Construction*, John Wiley, New York.

Kulhanek, O., 1990. *Anatomy of Seismograms*, Elsevier, Amsterdam.

Lamb, H., 1904. On the propagation of tremors over the surface of an elastic solid. *Phil. Trans. Roy. Soc.*, A, **203**, 1–42.

Langston, R. W., 1990. High-resolution refraction seismic data acquisition and interpretation. In Ward, S. H. (ed.), *Geotechnical and Environmental Geophysics*, Vol. 1, Soc. Explorat. Geophys., Tulsa, OK, pp. 45–74.

Lapwood, E. R. and Usami, T., 1981. *Free Oscillations of the Earth*, Cambridge University Press, Cambridge.

Latham, G., Ewing, M. and Dorman, J. *et al.*, 1971. Moonquakes. *Science*, **174**, 687–692.

Latter, J. H. (ed.), 1989. *Volcanic Hazards: Assessment and Monitoring*, Springer-Verlag, Berlin.

Lavergne, M., 1970. Emission by underwater explosions. *Geophysics*, **35**, 419–435.

Lay, T., 1987. Structure of the mantle and core. *Rev. Geophys.*, **25**, 1161–1167.

Lay, T., 1989. Structure of the core–mantle transition zone. *EOS*, **70** (4), 49–59.

Lay, T., 1994. Deep earth seismology. *Geotimes*, June, 12–15.

Lay, T., Ahrens, T. J., Olson, P., Smyth, J. and Loper, D., 1990. Studies of the Earth's deep interior: goals and trends. *Physics Today*, **43**, 44–52.

Leary, P. C., Crampin, S. and McEvilly, T. V., 1990. Seismic fracture anisotropy in the Earth's crust: an overview. *J. Geophys. Res.*, **95** (11), 105–114.

Lehmann, I., 1936. P[1]. *Pub. Bur. Cent. Seism. Int.*, A, **14**, 3–31.

Lehmann, I., 1964. On the travel times of P as determined from nuclear explosions. *Bull. Seism. Soc. Am.*, **54**, 123–139.

Lerner-Lam, A. L. and Jordan, T. H., 1987. How thick are the continents. *J. Geophs. Res.*, **92**, 14 007–14 026.

Leven, J. H., Finlayson, J. M. and Wright, C. (eds), 1987. Seismic probing of continents and their margins. *Tectonophysics*, **173**, 1–641.

Libbrecht, K. G. and Woodward, M. F., 1991. Advances in helioseismology. *Science*, **253**, 152–157.

Linde, A. T. and Silver, P. G., 1989. Elevation changes and the great 1960 Chilean earthquake — support for aseismic slip. *Geophys. Res. Lett.*, **16**, 1305–1308.

Lindh, A. G., 1983. US Geological Survey Open-File Report, 83/63.

Lines, L. R. and Treitel, S., 1984. Tutorial: a review of least squares inversion and its application to geophysical problems. *Geophys. Prospect.*, **32**, 159–186.

Lines, L. R., Tan, H. and Schultz, A. K., 1991. Cross-borehole analysis of velocity and density. *Geoexploration*, **28**, 183–191.

Lisowski, M., Savage, J. C. and Prescott, W. H., 1991. The velocity field along the San Andreas Fault in central and southern California. *J. Geophys. Res.*, **96**, 8369–8389.

Litak, R. K. and Brown, L. D., 1989. A modern perspective on the Conrad Discontinuity. *EOS, Trans. Am. Geophys. Un.*, 713–725.

Liu, L., 1985. *Elements, Oxides and Silicates: High Pressure Phases*, Oxford University Press, Oxford.

Liu, L., Zoback, M. D. and Segall, P., 1992. Rapid intraplate strain accumulation in the New Madrid seismic zone. *Science*, **257**, 1666–1669.

Lognonne, P. and Mosser, B., 1993. Planetary seismology. *Surveys in Geophysics*, **14**, 239–302.

Lomnitz, C., 1994, *Fundamentals of Earthquake Prediction*, John Wiley and Sons, New York.

Long, L. T., 1988. A model for major intracontinental earthquakes. *Seism. Res. Lett.*, **59**, 273–278.

Longuet-Higgins, M. S., 1950. A theory of the origin of microseisms. *Phil. Trans. Roy. Soc.*, A, **243**, 1–35.

Loper, D. E. and Lay, T., 1995. The core-mantle boundary region. *J. Geophys. Res.*, **100**, B4, 6397–6420.

Love, A. E. H., 1911. *Some Problems of Geodynamics*, Cambridge University Press, Cambridge.

Macelwane, J. B., 1940. Fifteen years of geophysics: a chapter in the exploration of the United States and Canada, 1924–1939. *Geophysics*, **5**, 250–258.

Maley, R. P., Porcella, R. and Etheridge, E., 1990. The strong motion network operated by the US Geol. Survey. *4th Nat. Conf. on Earthquake Engr.*, **1**, 347–356.

Mallet, R., 1848. *Irish Acad. Trans.*, **21**, 51–105.

Malone, S. D., Endo, E. T., Weaver, C. S. and Ramey, J. W., 1981. Seismic monitoring for eruption prediction. *US Geol. Surv. Paper 1250*, pp. 803–813.

Marillier, F., Hull, P., Roest, W. R. and Durling, P., 1990. New color display techniques help to interpret deep seismic sections. *EOS, Trans. Am. Geophys. Un.*, **71**, 1147–1149.

Masse, R. P. and Needham, R. E., 1989. NEIC – The National Earthquake Information Center. *Earthquakes and Volcanoes, USGS*, **21**, 4–44.

Massonet, D., Rossi, M., Carmona, C., Adragna, F., Peltzer, G., Feigl, K. and Rabaute, T., 1993. The displacement field of the Landers earthquake mapped by radar interferometry. *Nature*, **364**, 138–142.

Masters, T. G. and Shearer, P. M., 1990. Summary of seismological constraints on the structure of the Earth's core. *J. Geophys. Res.*, **95**, 21 691–21 695.

Masters, A., Jordan, T. H., Silver, P. G. and Gilbert, F., 1982. An aspherical Earth structure from fundamental spheroidal mode data. *Nature*, **298**, 609–613.

Mathews, D. H. and the BIRPS Group, 1990. Progress in BIRPS deep seismic profiling around the British Isles. *Tectonophysics*, **173**, 387–396.

Matsuzawa, A., Tamano, T., Aoki, Y. and Ikawa, T., 1980. Structure of the Japan Trench subduction zone from multi-channel seismic reflection records. *Marine Geology*, **35**, 171–182.

Maxwell, J. C., 1984. What is the lithosphere. *EOS, Trans. Am. Geophys. Un.*, 321–325.

Mayne, W. H., 1967. Practical considerations in the use of the common depth point techniques. *Geophysics*, **32**, 225–229.

McCaffrey, R. and Nabelek, J., 1984. The geometry of back arc thrusting along the eastern Sunda arc, Indonesia: constraints from earthquake and gravity data. *J. Geophys. Res.*, **89**, 6171–6179.

McCue, K., Wesson, V. and Gibson, G., 1990. The Newcastle, New South Wales, earthquake of 28 December 1989. *BMR J. of Austral. Geol. and Geophys.*, **11**, 559–567.

McGuire, R. K. and Arabasz, W. J., 1990. An introduction to probabilistic seismic data analysis. In Ward, S. H. (ed.), *Geotechnical and Environmental Geophysics*, Vol. 1, Soc. Explorat. Geophys., Tulsa, OK, pp. 333–353.

McNally, K. C., 1983. Seismic gaps in space and time. *Ann. Rev. Earth and Planet. Sci.*, **11**, 359–369.

McNally, K. and Ward, S. N., 1990. The Loma Prieta earthquake of October 17, 1989: introduction to the special issue. *Geophys. Res. Lett.*, **17**, 1177.

McNutt, S. R. and Sydnor, R. H., 1990. The Loma Prieta (Santa Cruz Mts), California, Earthquake of 17 October, 1989. California Dept. Conservation, Div. Mines and Geol., Special Pub. No. 104.

McQuillin, R., Bacon, M. and Barclay, W., 1979. *An Introduction to Seismic Interpretation*, Gulf Publishing, Houston, TX.

Meade, C. and Jeanloz, R., 1991. Deep-focus earthquakes and recycling of water into the Earth's mantle. *Science*, **252**, 68–72.

Meade, R. B., 1991. Reservoirs and earthquakes. *Engineering Geology*, **30**, 245–262.

Meissner, R., 1986. *The Continental Crust: A Geophysical Approach*, Academic Press, Orlando, FL.

Meissner, R. and Brown, L., 1991. Seismic reflections from the Earth's crust: comparative studies of tectonic patterns. *Geophys. J. Int.*, **105**, 1–2.

Meissner, R. and Mooney, W. D., 1991. Speculations on continental crustal evolution. *EOS, Trans. Am. Geophys. Un.*, **72**, 585–590.

Meissner, R. and Wever, T., 1992. The possible role of fluids for the structuring of the continental crust. *Earth Sci. Revs.*, **32**, 19–32.

Melton, B. S., 1981. Earthquake seismograph development: a modern history – Parts 1 and 2. *EOS, Trans. Am. Geophys. Un.*, **62** (21), 505–510; (25), 545–548.

Menke, W., 1984. *Geophysical Data Analysis: Discrete Inverse Theory*, Academic Press, FL.

Menzies, M. A. (ed.), 1990. *Continental Mantle*, Clarendon, Oxford.

Mereu, R. F., Mueller, S. and Fountain, D. M. (eds), 1989. *Properties and Processes of the Earth's Lower Crust*, Geophys. Monograph 51, Am. Geophys. Union.

Miller, R. D. and Steeples, D. W., 1991. Detecting voids in a 0.6 m coal seam, 7 m deep, using seismic reflection. *Geoexploration*, **28**, 109–119.

Minoura, K. and Hasegawa, A., 1992. Crustal structure and origin of the northeast Japan arc. *The Island Arc*, **1**, 2–15.

Minster, J. B. and Jordan, T. H., 1978. Present day plate motions. *J. Geophys. Res.*, **83**, 5331–5354.

Minster, J. B. and Jordan, T. H., 1987. Vector constraints on western US deformation from space geodesy, neotectonics and plate motions. *J. Geophys. Res.*, **92**(B6), 4798–4804.

Molnar, P. and Chen, W.-P., 1982. Seismicity and mountain building. In Hsu, K. G. (ed.), *Mountain Building Processes*, Academic Press, London.

Molnar, P. and Lyon-Caen, H., 1989. Fault plane solutions of earthquakes and active tectonics of the Tibetan plateau and its margins. *Geophys. J. Int.*, **99**, 123–153.

Montagne, J.-P. and Tanimoto, T., 1991. Global upper mantle tomography of seismic velocities and anisotropies. *J. Geophys. Res.*, **96**(B12), 337–351.

Mooney, W. D., 1989. Seismic methods for determining earthquake source parameters and lithospheric structure. In Pakiser, L. C. and Mooney, W. D. (eds), *Geophysical Framework of the Continental United States*, Geol. Soc. Amer. Mem. No. 172, pp. 11–33.

Mooney, W. D. and Meissner, R., 1991, Continental crustal evolution. *EOS*, **72**, 537–541.

Mooney, W. D. and Meissner, R., 1992. Multi-genetic origin of crustal reflectivity: a review of seismic reflection profiling of the continental lower crust and Moho. In Fountain, D. M., Arculus, R. and Kay, R. W. (eds), *Continental Lower Crust*, Elsevier, Amsterdam, pp. 45–79.

Morelli, A., Dziewonski, A. M. and Woodhouse, J. H., 1986. Anisotropy of the inner core inferred from PKIKP travel times. *Geophys. Res. Lett.*, **13**, 1545–1548.

Morgan, J. P. and Chen, Y. J., 1993. The genesis of oceanic crust: magma injection, hydrothermal circulation, and crustal flow. *J. Geophys. Res.*, **98**, 6283–6297.

Murphy, A. J. and Savino, J. M., 1975. A comprehensive study of long-period (20–200 sec) earth noise at the high-gain world wide seismograph stations. *Bull. Seism. Soc. Am.*, **65**, 1827–1862.

Musgrave, A. W. (ed.), 1967. *Seismic Refraction Surveying*, Soc. Explorat. Geophysicists, Tulsa, OK.

Mutter, J. C., 1986. Seismic images of plate boundaries. *Scientific American*, **254**, 54–61.

Mutter, J. C. and Karson, J. A., 1992. Structural processes at slow spreading ridges. *Science*, **257**, 627–637

Nafe, J. E. and Drake, C. L., 1963. Physical properties of marine sediments. In Hill, M. N. (ed.), *The Sea*, Vol. 3, Interscience, New York, pp. 794–815.

Nataf, H. C., Nakanishi, I. and Anderson, D. L., 1986. Measurements of mantle wave velocities and inversion of lateral heterogeneities and anisotropy. 3. Inversion. *J. Geophys. Res.*, **91**, 7261–7307.

Neidell, N. S. and Poggiagliolmi, F., 1977. Stratigrophic modelling and interpretation. In Payton, C. E. (ed.). *Seismic Stratigraphy Applications in Hydrocarbon Exploration.* Amer. Assoc. Petrol. Memoir 26, pp. 389–416.

Newton, R. C., 1990. Fluids and shear zones in the deep crust. *Tectonophysics*, **182** (1/2), Special Issue on the Nature of the Lower Crust.

Ni, J. and Barazangi, M., 1984. Seismotectonics of the Himalayan collision zone: geometry of the underthrusting Indian plate beneath the Himalaya. *J. Geophys. Res.*, **89**, 1147–1163.

Niazi, M. and Anderson, D. L., 1965. Upper mantle structure in western North America from apparent velocities of P waves. *J. Geophys. Res.*, **70**, 4633–4640.

Nishenko, S. P. and Bollinger, G. A., 1990. Forecasting earthquakes in the central and eastern United States. *Science*, **249**, 1412–1416.

Nolet, G. (ed.), 1987. *Seismic Tomography – With Applications in Global Seismology and Exploration Geophysics*, Reidel, Dordrecht.

ODP Leg 148 Shipboard Scientific Party, 1993. ODP Leg 148 barely misses deepest layer. *EOS*, **74**, 489.

Oldham, R. D., 1906. The constitution of the interior of the earth as revealed by earthquakes. *Quart. J. Geol. Soc. London*, **62**, 456–475.

Oliver, J., 1960. Long earthquake waves. *Scientific American*, p. 77.

Oliver, J., 1988. Opinion. *Geology*, **16**, 291.

Olsen, P., Silver, P.G. and Carlson, R. W., 1990. The large-scale structure of convection in the Earth's mantle. *Nature*, **344**, 209–215.

Oral, M. B., Reilinger, R. E., Toksoz, M. N. *et al.* 1995. Global positioning system offers evidence of plate motions in Eastern Mediterranean. *EOS*, **76**, 9–11.

Ostrander, W. J., 1984. Plane-wave reflection coefficients for gas sands — sands at non-normal angles of incidence. *Geophysics.*, **49**, 1637–1648.

Pacheco, J. F. and Sykes, L. R., 1992. Seismic moment catalogue of large shallow earthquakes, 1900 to 1989. *Bull. Seism. Soc.*, **82**, 1306–1349.

Pacheco, J. F., Scholz, C. H. and Sykes, L. R., 1992. Changes in frequency–size relationship from small to large earthquakes. *Nature*, **355**, 71–73.

Pakiser, L. C. and Mooney, W. D., 1989. Geophysical Framework of the Continental United States. *Geol. Soc. Am. Mem.* 172.

Palmer, D., 1990. The generalized reciprocal method — an integrated approach to shallow refraction seismology. *Explorat. Geophys.*, **21**, 33–44.

Panel on Seismic Hazard Analysis, 1988. *Probabilistic Seismic Hazard Analysis*, Nat. Acad. Press, Washington, DC.

Parasnis, D. S., 1986. *Principles of Applied Geophysics*, 4th edn, Chapman and Hall, London.

Park, S. K., Johnston, M. J. S., Maddern, T. R., Morgan, F. D., Morrison, H. F., 1993. Electromagnetic precursors to earthquakes in the ULF band: a review of observations and mechanisms. *Rev. Geophys.*, **31**(2), 117–132.

Parson, L. M., Murton, B. J. and Browning, P. (eds), 1992. *Ophiolites and Their Modern Oceanic Analogues*, Geol. Soc. Special Pub. No. 60.

Paulis, G. L., 1989. Inverse theory. In James, D. E. (ed.) *The Encyclopedia of Solid Earth Geophysics*, Van Nostrand Reinhold, New York, pp. 647–654.

Pelayo, A. M. and Wiens, D. A., 1992. Tsunami earthquakes: slow thrust-faulting events in the accretionary wedge. *J. Geophys. Res.*, **97**, 15 321–15 337.

Peltier, W. R. (ed.), 1989. *Mantle Convection: Plate Tectonics and Global Dynamics*, Gordon and Breach, New York.

Percival, J. A. and Berry, M. J., 1987. The lower crust of

the continents. In Fuchs, K. *et al.* (*eds*), *Composition, Structure and Dynamics of the Lithosphere–Asthenosphere System.* Geodynamics Series, Amer. Geophys. Union, Vol. 16., pp. 33–60.

Peterson, J. and Orsini, N. A., 1976. Seismic research observations: Upgrading the worldwide seismic data network. *EOS, Trans. Am. Geophys. Un.*, **57**, 548–556.

Phillips, W. C. and Fehler, M. C., 1991. Traveltime tomography: a comparison of popular methods. *Geophysics*, **56**, 1639–1649.

Pilant, W. L., 1979. *Elastic Waves in the Earth*, Elsevier, Amsterdam.

Plafker, G., 1969. Tectonics of the March 27, 1964 Alaska earthquake. US Geol. Surv. Prof. Paper 543-1, pp. 1–74.

Plafker, G., 1972. Alaskan earthquake of 1964 and Chilean earthquake of 1960: implications for arc tectonics. *J. Geophys. Res.*, **77**, 901–925.

Powell, R. E. and Weldon, R. J., 1992. Evolution of the San Andreas Fault. *Ann. Rev. Earth and Planet. Sci.*, **20**, 431–468.

Press, F., 1965. Resonant vibrations of the earth. *Scientific American*, Nov., p. 52.

Press, F. and Siever, R., 1986. *Earth*, 3rd edn, W. H. Freeman, San Francisco.

Pullan, S. E. and Hunter, J. A. 1990. Delimitation of buried bedrock valleys using the optimum offset shallow seismic reflection technique. In Ward, S. H. (ed.), *Geotechnical and Environmental Geophysics*, Vol. 3, Soc. Explorat. Geophys., Tulsa, OK, pp. 75–88.

Rayleigh, Lord (Strutt, J. W.), 1885. On waves propagated along the plane surface of an elastic solid. *Proc. London Math. Soc.*, **17**, 4–11.

Rector, J. W. and Marion, B. P., 1991. The use of drill-bit energy as a downhole seismic source. *Geophysics*, **56**, 628–634.

Reid, H. F., 1911. The elastic-rebound theory of earthquakes. *Bull. Dept., Geol. Univ. Cal.*, **6**, 413–444.

Reshef, M. and Kosloff, D., 1986. Migration of common-shot gathers. *Geophysics.*, **51**, 324–331.

Revenaugh, J. and Jordan, T. H., 1991. Mantle layering from ScS reverberations. *J. Geophys. Res.*, **96**(B12), 19 749–19 824.

Ribe, N. M., 1989. Seismic anisotropy and mantle flow. *J. Geophys. Res.*, **94**, 4213–4223.

Richards, P. G. and Zavales, J., 1990. Seismic discrimination of nuclear explosions. *Ann. Rev. Earth and Planet. Sci.*, **18**, 257–286.

Richards, T. C., 1961. Motion of the ground on arrival of reflected longitudinal and transverse waves at wide-angle reflection distances. *Geophysics*, **26**, 277–297.

Richter, C. F., 1935. An instrumental earthquake magnitude scale. *Bull. Seism. Soc. Am.*, **25**, 1–32.

Richter, C. F., 1958. *Elementary Seismology*, Freeman, San Francisco.

Rikitake, T., 1982. *Earthquake Forecasting and Warning*, Reidel, Dordrecht.

Rind, D. and Donn, W., 1978. Microseisms at Palisades 1, source location and propagation. *J. Geophys. Res.*, **83**, 3525–3534.

Ringdal, F. (ed.), 1990. Regional seismic arrays and nuclear test ban verification. *Bull. Seism. Soc. Am.*, **80**(B), Special Symposium Issue.

Ringwood, A. E., 1975. *Composition and Petrology of the Earth's Mantle*, McGraw-Hill, New York.

Ringwood, A. E., 1983. The Bakerian Lecture. The Earth's core: its composition, formation and bearing on the origin of the Earth. *Proc. Roy. Soc. London*, A, **395**, 1–46.

Robinson, E. A., 1983a. *Seismic Velocity Analysis and the Convolutional Model*, Reidel, Boston, MA.

Robinson, E. A., 1983b. *Migration of Geophysical Data*, Int. Human Res. Dev. Corp., Boston, MA.

Robinson, E. A. and Treitel, S., 1980. *Geophysical Signal Analysis*, Prentice-Hall, Englewood Cliffs, NJ.

Robinson, E. S. and Coruh, C., 1988. *Basic Exploration Geophysics*, John Wiley, New York.

Roeloffs, E. and Langbein, J., 1994. The earthquake prediction experiment at Parkfield, California. *Rev. Geophys.*, **32**(3), 315–336.

Romanowicz, B., 1991. Seismic tomography of the Earth's mantle. *Ann. Rev. Earth and Planet. Sci.*, **19**, 77–99.

Romanowicz, B. A. and Dziewonski, A. M., 1987. Global digital seismographic network: research opportunities and recent initiatives. *Am. Geophys. Union, Geodynamics Series*, **16**, 99–110.

Romig, P. R. (ed.), 1986. Special issue on engineering and groundwater geophysics. *Geophysics*, **51**, 221–323.

Rundle, J. B., Elbring, C. J., Striker, R. P., *et al.*, 1985. Seismic imaging in Long Valley, California, by surface and borehole techniques: an investigation of active tectonics. *EOS, Trans. Am. Geophys. Un.*, **66**(18), 194–201.

Salisbury, M. H. and Fountain, D. M., 1990. *Exposed Cross-Sections of the Continental Crust*, Kluwer Academic Publishers, Dordrecht.

Sato, K., 1993. Tectonic plate motion and deformation inferred from very long baseline interferometry. *Tectonophysics*, **220**, 69–87.

Savage, J. C. and Lisowski, M., 1991. Strain measurements and the potential for a great subduction earthquake off the coast of Washington. *Science*, **252**, 101–103.

Savage, J. C., Lisowski, M. and Prescott, W. H., 1986. Strain accumulation in the Shumagin and Yakataga seismic gaps, Alaska. *Science*, **231**, 585–587.

Scheimer, J. F. and Borg, I. Y., 1984. Deep seismic sounding with nuclear explosives in the Soviet Union. *Science*, **226**, 787–792.

Schneider, W. A., 1978. Integral formulation for migration in two dimensions. *Geophysics*, **43**, 49–76.

Scholz, C. H., 1990. *The Mechanics of Earthquakes and Faulting*, Cambridge University Press, Cambridge.

Scholz, C. H., Aviles, C. A. and Wesnousky, S. G., 1986. Scaling differences between large interplate and intraplate earthquakes. *Bull. Seism. Soc. Am.*, **76**, 65–70.

Schwartz, D. P., 1987. Earthquakes of the Holocene. *Rev. Geophysics*, **25**(6), 1197–1202.

Sengbush, R. L., 1983. *Seismic Exploration Methods*, Int. Human Res. Dev. Corp., Boston, MA.

Sharma, P. V., 1986. *Geophysical Methods in Geology*, 2nd edn, Elsevier, New York.

Shaw, B. E., Carlson, J. M. and Langer, J. S., 1992. Patterns of seismic activity preceding large earthquakes. *J. Geophys. Res.*, **97**(B1), 479–488.

Shearer, P. M., 1990. Seismic imaging of upper-mantle structure with new evidence for a 520km discontinuity. *Nature*, **344**, 121–125.

Sheriff, R. E., 1980. *Seismic Stratigraphy*, Int. Human Resour. Dev. Corp., Boston, MA.

Sheriff, R. E., 1984. *Encyclopedic Dictionary of Exploration Geophysics*, 2nd edn, Soc. Explor. Geophys., Tulsa, OK.

Sheriff, R. E. and Geldart, L. P., 1982. *Exploration Seismology*, Vol. 1, *History, Theory and Data Acquisition*, Cambridge University Press, Cambridge.

Sheriff, R. E. and Geldart, L. P., 1983. *Exploration Seismology*, Vol. 2, *Data Processing and Interpretation*, Cambridge University Press, Cambridge.

Shuey. R. T., 1985. A simplification of Zoeppritz equations. *Geophysics.*, **50**, 609–614.

Sibson, R. H., 1982. Fault zone models, heat flow, and the depth distribution of earthquakes in the continental crust. *Bull. Seism. Soc. Am.*, **72**, 151–163.

Sibson, R. H., 1986. Earthquakes and lineament infrastructure. *Phil. Trans. Roy. Soc. London*, A, **317**, 63–79.

Sibson, R. H., 1989. Earthquake faulting as a structural process. *J. Struct. Geol.*, **11**, 1–14.

Sieh, K., Stuiver, M. and Brillinger, D., 1989. A more precise chronology of earthquakes produced by the San Andreas fault in southern California. *J. Geophys. Res.*, **94**(B1), 603–623.

Silvia, M. T. and Robinson, E. A., 1979. *Deconvolution of Geophysical Time Series in the Exploration for Oil and Natural Gas*, Elsevier, Amsterdam.

Simon, R. B., 1981. *Earthquake Interpretations, A Manual for Reading Seismograms*, W. Kaufmann, Los Altos, CA.

Simpson, D. W. and Richards, P. G. (eds), 1981. *Earthquake Prediction*, Maurice Ewing Series, IV, Am. Geophys. Union, Washington, DC.

Singh, S. K., Mena, E. and Castro, R., 1988. Some aspects of the source characteristics of the 19 September 1985 Michoacan earthquake and ground motion amplification in and near Mexico City from strong motion data. *Bull. Seism. Soc Ames.*, **78**, 451–477.

Sinton, J. M. and Detrick, R. S., 1992. Mid-ocean ridge chambers. *J. Geophys. Res.*, **97**(B1), 197–216.

Sjogren, B., 1984. *Shallow Refraction Seismics*. Chapman and Hall, London.

Smith, D. G. (ed.), 1982. *The Cambridge Encyclopaedia of Earth Science*, Cambridge University Press, Cambridge.

Smith, S. W., 1986. IRIS: a program for the next decade. *EOS, Trans. Am. Geophys. Un.*, **67**, 213–219.

Smith, W. D., 1986. Evidence for precursory changes in the frequency–magnitude b value. *Geophys. J. Roy. Astron. Soc.*, **86**, 815–838.

Souriau, A. and Woodhouse, J. H., 1985. A worldwide comparison of predicted S wave delays from a three dimensional upper mantle model with P wave station corrections. *Phys. Earth and Planet. Int.*, **39**, 75–88.

Stacey, F. D., 1977. *Physics of the Earth*, 2nd edn, John Wiley and Sons, New York; 1992, 3rd edn, Brookfield Press, Brisbane.

Stanley, M. and Singh, R., 1991. Supercomputers in seismic data processing. *Explor. Geophys.*, **22**, 379–382.

Steeples, D. W. and Miller, R. D., 1990. Seismic reflection methods applied to engineering, environmental and groundwater problems. In Ward, S. H. (ed.), *Geotechnical and Environmental Geophysics*, Vol. 1, Soc. Explorat. Geophys., Tulsa, OK, pp. 1–30.

Stein, M. and Hoffmann, M., 1994. Mantle plumes and episodic crustal growth. *Nature*, **372**, 63–67.

Stoffa, P. L. (ed.), 1989. *Tau-P: A Plane Wave Approach to the Analysis of Seismic Data*, Kluwer Academic, Dordrecht.

Stolt, R. H. and Benson, A. K., 1986. *Seismic Migration*, Geophysical Press, London.

Stoneley, R., 1967. History of modern seismology. In Runcorn, S. K. (ed.), *International Dictionary of Geophysics*, Pergamon, Oxford, pp. 724–729.

Stover, C. W. and Coffman, J. L., 1993. *Seismicity of the United States*, 1568–1989 (*revised*). US Geolog. Survey Paper 1527.

Stump, B. W., 1991. Nuclear explosion seismology: verification, source theory, wave propagation and politics. *Rev. Geophys.*, Suppl. US National Rep. to IUGG, pp 734–741.

Sykes, L. R., 1971. Aftershock zones of great earthquakes, seismicity gaps and earthquake prediction for Alaska and the Aleutians. *J. Geophys. Res.*, **75**, 8021–8041.

Sykes, L. R., 1978. Intra-plate seismicity, reactivation of pre-existing zones of weakness, alkaline magmatism and other tectonism post-dating continental fragmentation. *Rev. Geophys. and Space Sci.*, **16**, 621–688.

Sykes, L. R. 1983. Predicting great earthquakes. In E. Fermi School of Physics Proceedings, *Earthquakes, Observations, Theory and Interpretation*. H. Kananori (ed.), North-Holland, Amsterdam, pp. 398–435.

Sykes, L. R., 1987. Underground nuclear tests: verifying limits on underground testing, yield estimates and public policy. *Rev. Geophys.*, **25**, 1209–1214.

Sykes, L. R. and Seeber, L., 1985. Great earthquakes and great asperities, San Andreas fault, California. *Geology*, **13**, 835–838.

Takemoto, S., 1991. Some problems on detection of earthquake precursors by means of continuous monitoring of crustal strains and tilts. *J. Geophys. Res.*, **96** (B6), 10 377–10 390.

Talwani, M., 1989. Ocean–continent transition. In James, D. E. (ed.), *Encyclopedia of Solid Earth Geophysics*, Van Nostrand Reinhold, New York, pp. 858–871.

Talwani, M., Stoffa, P., Buhl, P., Windisch, C. and Diebold, J. B., 1982. Seismic multichannel towed arrays in the exploration of the oceanic crust. *Tectonophysics*, **81**, 273–300.

Taner, M. T. and Koehler, K. F., 1969. Velocity spectra: Digital computer derivation and applications of velocity functions. *Geophysics*, **34**, 859–881.

Tanioka, Y., Ruff, L. and Satake, K., 1993. The 1993 Japan Sea earthquake. *EOS*, **74**, 377–380.

Tate, J. and Daily, W., 1989. Evidence of electro-seismic phenomena. *Phys. Earth and Planet. Int.*, **57**, 1–10.

Telford, W. M., Geldart, L. P., Sheriff, R. E. and Keys, D. A., 1990. *Applied Geophysics*, 2nd edn, Cambridge University Press, London.

Thatcher, W., 1984. Earthquake deformation cycle, recurrence and the time-predictable model. *J. Geophys. Res.*, **89**, 5674–5680.

Thatcher, W., 1990. Order and diversity in the modes of circum-Pacific earthquake recurrence. *J. Geophys. Res.*, **95**, 2609–2623.

Thomson, W., 1863. On the rigidity of the earth. *Phil. Trans. Roy. Soc. London*, **153**, 573.

Thurber, C. H. and Aki, K., 1987. Three-dimensional seismic imaging. *Ann. Rev. Earth and Planet. Sci.*, **15**, 115–139.

Tilling, R. I. (ed.), 1989. *Volcanic Hazards*, Amer. Geophys. Union, Washington, DC.

Toksoz, M. N. and Johnston, D. H. (eds), 1981. *Seismic Wave Attenuation*, Soc. Explor. Geophys., Tulsa, OK.

Toksoz, M. N., Goins, N. R. and Cheng, C. H., 1977. Moonquakes: mechanisms and relation to tidal stresses. *Science*, **196**, 979–981.

Trehu, A. M., Ballard, A., Dorman, L. M., Gettrust, J. F. and Schreiner, A., 1989. Structure of the lower crust beneath the Caroline Trough, US Atlantic continental margin. *J. Geophys. Res.*, **94**, 10 585–10 600.

Trifunac, M. D. and Brady, A. G., 1975. On the correlation of seismic intensity scales with the peaks of recorded strong ground motion. *Bull. Seism. Soc. Am.*, **65**, 139–162.

Trifunac, M. D. and Brady, A. G., 1976. Correlations of peak acceleration, velocity and displacement with earthquake magnitude, distance and site conditions. *Earthquake Eng. and Struct. Dynamics*, **4**, 455–471.

Trorey, A. W., 1970. A simple theory for seismic diffractions. *Geophysics*, **35**, 762–784.

Tsapanos, T. M. and Burton, P. W., 1991. Seismic hazard evaluation for specific seismic regions of the world. *Tectonophysics*, **194**, 153–169.

Tucker, W., Herrin, E. and Freedman, H. W., 1968. Some statistical aspects of the estimation of seismic travel times. *Bull. Seismol Soc. Amer.*, **58**, 1243–1261.

Turcotte, D. L., 1991. Earthquake prediction. *Ann. Rev. Earth and Planet. Sci.*, **19**, 263–281.

Ursin, B. and Dahl, T., 1992. Seismic reflection amplitudes. *Geophys. Prospect.*, **40**, 483–512.

US Geological Survey Staff, 1990. The Loma Prieta, California earthquake: an anticipated event. *Science*, **247**, 286–293.

US Geological Survey, 1993. Earthquakes and volcanoes. USGS, **24**, 6.

US Geological Survey Scientists and the Southern California Earthquake Center, 1994. The magnitude 6.7 Northridge, California, earthquake of January, 1994. *Science*, **266**, 389–397.

Uyeda, S., 1978. *The New View of the Earth*, Freeman and Co.

Vail, P. R., Todd, R. G. and Sangree, J. B., 1977. Chronostratigraphic significance of seismic reflections. In Payton, C. E. (ed.), Seismic Stratigraphy — Applications to Hydrocarbon Exploration. *Am. Assoc. Petrol. Geol. Mem.*, **26**, 99–116.

Varotsos, P. and Kulhanek, O. (eds), 1993. Measurements and theoretical models of the Earth's electric field variations related to earthquakes. *Tectonophysics*, **224** (1/3) (special issue).

Vaughan, P. S., 1993. North-West Shelf 3-D seismic — a decade of developments. *APEA* (*Aust. Petr. Explor. Assoc.*) *J.*, **93**, 315–321.

Vita-Finzi, C., 1986. *Recent Earth Movements*, Academic Press, London.

Vittori, E., Labini, S. S. and Serva, L., 1991. Palaeoseismology: a review of the state-of-the-art. *Tectonophysics*, **193**, 9–32.

Vogel, G., Gorenflo, R., Kummer, B. and Ofoegbu, C. O. (eds), 1988. *Inverse Modelling in Exploration Geophysics*, Vieweg. Braunschweig/Wiesbaden.

Wakita, H., Nakamura, Y. and Sano, Y., 1988. Short-term and intermediate-term geochemical precursors. *Pure and Appl. Geophys.*, **126**, 267–278.

Walker, C., Leung, T. M., Win, M. A. and Whiteley, R. J., 1991. Engineering seismic refraction: an improved field practice and a new interpretation program, *Explor. Geophysics*, **22**, 423–428.

Walker, D. A., 1984. Deep ocean seismology. *EOS, Trans. Am. Geophys. Un.*, **65** (1), 2–3.

Wallace, R. E. (ed.), 1990. *The San Andreas Fault System, California*, US Geol. Surv. Profess. Paper 1515.

Ward, S. H., 1990. *Geotechnical and Environmental Geophysics*, 3 vols, Soc. Explorat. Geophysicists, Tulsa, OK.

Ward, S. N., 1989. Tsunamis. In James, D. E. (ed.). *The Encyclopedia of Solid Earth Geophysics*, Van Nostrand Reinhold, New York, pp. 1279–1292.

Wardel, J. and Whiting, P., 1989. Multiple attenuation: some current techniques. *Explor. Geophys.*, **20**, 275–279.

Warner, M., 1990. Basalts, water, or shear zones in the lower continental crust. *Tectonophysics*, **173**, 163–174.

Waters, K. H., 1987. *Reflection Seismology, A Tool for Energy Resource Exploration*, 3rd edn, John Wiley, New York.

Weatherby, B. B., 1940. The history and development of

seismic prospecting. *Geophysics*, **5**, 215.

Weaver, C. S. and Malone, S. D., 1987. Overview of the tectonic setting and recent studies of eruptions of Mt St Helens, Washington. *J. Geophys. Res.*, **92**, 10 149–10 154.

Werth, G. C., Liu, D. T. and Trorey, A. W., 1959. Offshore singing — field experiments and theoretical interpretation. *Geophysics.*, **24**, 220–232.

Wesson, R. L. and Wallace, R. E., 1985. Predicting the next great earthquake in California. *Scientific American*, **252**, 23–31.

White, J. E., 1983. *Underground Sound, Application of Seismic Waves*, Elsevier, Amsterdam.

White, R. S. and McKenzie, D. P., 1989. Volcanism at rifts. *Scientific American*, **261**, 44–55.

Whitmarsh, R. B., Bott, M. H. P., Fairhead, J. D. and Kuznir, N. J., *et al.* (eds), 1991. Tectonic stress in the lithosphere. *Phil. Trans. Roy. Soc.*, **337**, 1–194.

Wiechert, E., 1896. *Verh. Ges. Deutsch Naturforsch. Aerzte*, **2**, 42.

Wiechert, E., 1910. Bestimmung des Weges der Erdbebenwellen im Erdinnen. *Physikalische Zeitschrift*, **11**, 249–310.

Wilson, M., 1993. Plate-moving mechanisms: constraints and controversies. *J. Geol. Soc., London*, **150**, 923–926.

Woodhouse, J. H. and Dziewonski, A. M., 1984. Mapping the upper mantle: Three dimensional modelling of earth structure by inversion of seismic waveforms. *J. Geophys. Res.*, **89**, 5953–5986.

Woods, B. B. and Helmberger, D. V., 1993. A new seismic discriminant for earthquakes and explosions. *EOS, Trans. Am. Geophys. Un.*, **74**(8), 91.

Working Group on California Earthquake Probabilities, 1995. *Bull. Seismol. Soc. Amer.*, **85** 2, 379–439.

Wright, C. and Cleary, J. R., 1972. P wave travel-time gradient measurements for the Warramunga seismic array and lower mantle structure. *Phys. Earth and Planet. Interiors.*, **5**, 213–230.

Wright, T. L. and Pierson, T. C., 1992. *Living With Volcanoes*, US Geological Survey, Circular 1073.

Wyss, M. (ed.), 1991. *Evaluation of Proposed Earthquake Precursors*, Am. Geophys. Union, Washington.

Wyss, M. and Habermann, R. E., 1988. Precursory seismic quiescence. *Pageoph.*, **126**, 319–331.

Wyss, M., Habermann, R. E. and Bodin, P., 1992. Seismic quiescence: a test of the hypothesis and a precursor to the next Parkfield, California, earthquake. *Geophys. J.*, **110**, 518–536.

Yeh, H., Imamura, F., Synolakis, C., Tsuji, Y., Liu, P. and Shi, S., 1993. The Flores Island tsunamis. *EOS.*, **74**, 369–373.

Yilmaz, O., 1987. *Seismic Data Processing*, Soc. Explor. Geophysicists, Tulsa, OK.

Yilmaz, O. and Claerbout, J. F., 1980. Prestack partial migration. *Geophys.*, **45**, 1753–1777.

Young, C. J. and Lay, T., 1987. The core–mantle boundary. *Ann. Rev. Earth and Planet. Sci.*, **15**, 25–46.

Young, J. B., Lilwell, R. C. and Douglas, A., 1988. *World Seismicity Maps Suitable for the Study of Seismic and Geographical Regionalization*, Atomic Weapons Research Establishment Report 07/87, HMSO, London.

Stuart, W. D. and Aki, K. (eds), 1988. Intermediate-term earthquake prediction. *Pageoph.*, **126**(2–4), 175–718.

Zelt, C. A. and Smith, R. B., 1992. Seismic travel-time inversion for 2D crustal velocity structure. *Geophys. J. Int.*, **108**, 16–34.

Zhao, W., Nelson, K. D. and Project Indepth Team, 1993. Deep seismic reflection evidence for continental underthrusting beneath southern Tibet. *Nature*, **366**, 557–559.

Zhou, H., Allen, C. R. and Kanamori, H., 1983. Rupture complexity of the 1970 Tongai and 1973 Luhuo earthquakes, China, from P-wave inversion, and the relationship to surface faulting. *Bull. Seism. Soc. Am.*, **73**, 1585–1597.

Ziolkowski, A., 1979. Seismic profiling for coal on land. In Fitch, A. A. (ed.), *Developments in Geophysical Exploration Methods – 1*, Applied Science Publishers, London, pp. 271–306.

Zoback, M. D. and Lachenbruch, A. H., 1992. Introduction to special section on the Cajon Pass scientific drilling project. *J. Geophys. Res.*, **97** (B4), 4991–4994.

Zoback, M. L., 1992. First and second order patterns of stress in the lithosphere: the world stress map project. *J. Geophys. Res.*, **97**(B8), 11 703–11 728.

Zoback, M. L. and Zoback, M. D., 1989. Tectonic stress field of the continental United States. In Pakiser, L. C. and Mooney, W. D. (eds), *Geophysical Framework of the Continental United States*, Geol. Soc. Amer. Mem. 172, pp. 523–539.

Index